T0134417

Making Healthcare Green

Nina S. Godbole • John P. Lamb

Making Healthcare Green

The Role of Cloud, Green IT, and Data Science to Reduce Healthcare Costs and Combat Climate Change

Nina S. Godbole
Researcher in Healthcare & Green IT
Certified Green IT Professional, Certified
Information Privacy Professional (CIPP/IT),
IAPP (USA)
Pune, India

John P. Lamb
Math Department
Pace University
Pleasantville, NY, USA

Author of "The Greening of IT, How
Companies Can Make a Difference for the
Environment"

ISBN 978-3-030-07718-1   ISBN 978-3-319-79069-5   (eBook)
https://doi.org/10.1007/978-3-319-79069-5

Printed on acid-free paper

This Springer imprint is published by the registered company Springer International Publishing AG part
of Springer Nature.
The registered company address is: Gewerbestrasse 11, 6330 Cham, Switzerland

*John Lamb dedication: For Penny, Jack,
Erin, and Keaton*

*Nina Godbole dedication: For Keya, Sudhir,
Ketaki, and my parents*

Foreword

We, the human race, are very interesting. We innovate, explore, and develop new technologies, economic models, political systems, and a lot more! And then we innovate and explore to solve the problems created by these creations of ours. The environment and its ensuing sustainability challenges are the results of our successes in technology and economic development. The very first steam engine in the mid-1800s was led by a man with a red flag to control its speed and reduce pollution! In the same era, Charles Babbage produced the first digitally programmable computing machine. He had not the wildest idea that his invention would eventually lead to massive data servers that churn megatons of carbon.

This age of Big Data is resulting in an incomprehensible explosion in the digital space. This is a dual-edged sword. Predictive Analytics, Machine Learning, IoT, and Cloud all offer strategic utilization of Big Data Technologies and Analytics to provide value to businesses. At the same time, ever-increasing use of these digital technologies threatens to swamp us with electronic wastage. Technology, for example, has made the tasks of finding blockages in arteries easy. And millions of MRI scans of the brain and lungs provide diagnosis of cancer that could not be detected half a century ago with the most sophisticated instruments of that time. This technology usage, however, generates carbon and electronic wastage in mind-boggling proportions. There is an acute need to undertake a delicate balancing act: use technology to save lives and, at the same time, control carbon and wastage resulting from the effort.

This is precisely the key contribution of this excellent book by Godbole and Lamb. The authors have investigated in detail the environmental challenges of running hospitals which form half the healthcare sector. They visited hospitals, interviewed decision-makers, and observed the myriad business processes and systems around hospitals. As a result, their book is replete with excellent, practical discussions on "Making Healthcare Green" by synergizing two major challenges of the day – health and environment.

The environment is not the preserve of the greenies, and sustainability is no longer a fringe issue. Interest in greening of healthcare is the natural outcome of the world's interest in carbon emissions and electronic wastages. A simple understanding of sustainability is that we ensure that any enterprise does not put more carbon and wastage in the environment than the bare necessity to achieve its business goals. Healthcare, in particular, has a major impact on the world's carbon footprint – ranging from 7% to 9% of the overall carbon generated by the world. At the same time, even small optimizations in the healthcare processes promise to impact their carbon footprints. For example, the use of EMRs (electronic medical records) has the potential to reduce carbon dioxide emissions by as much as 1.7 million tons across the United States and also reduce the use of 1044 tons of paper for medical charts annually.

This book addresses these issues in a holistic and pragmatic way. The authors bring in Big Data Technologies associated with Data Science and Analytics and the use of Internet-of-Things (IoT) devices in improving the quality of healthcare and reducing its carbon impact. The authors take great care to keep their discussions in the context of corporate social responsibilities of organizations in a market-driven economy. Thus, they nicely tie the need for greening of healthcare with the obvious advantages of greater efficiency. Their work reflects what I wrote in the preface of my own book a few years ago:

> Profits versus carbon, customer services versus carbon, competition versus carbon, regulations versus carbon. For too long the carbon reduction debate has pitched good environmental outcomes against good business outcomes. Yet the reality, however, is that best business practice delivers both good business outcomes and environmental benefits. Many CEOs are either looking to show leadership or are leading in carbon reduction because it is good for their business. (Green IT Strategies and Applications: Using Environmental Intelligence, CRC Press, USA, 2011)

This book presents practical information on how green healthcare (whose major sub-segment is "hospitals") can contribute to solving the climate crisis by adopting green IT practices. Green healthcare, green computing, and green IT (information technology) are all excellent ways to reduce electricity use and save money doing it.

I congratulate the authors for a thorough job; they both have to their credit their contributions in the green IT domain – Godbole contributed chapters to the *Handbook of Research in Green ICT: Technology, Business, and Social Perspectives*, IGI, and was a member of the review team for the book *Harnessing Green IT: Principles and Practices*, while Dr. Lamb is the author of the book *The Greening of IT*. I wish you, the readers, my very best in reading, understanding, and applying the nuggets presented in a lucid and practical way in this book.

Sarasota, FL, USA   Bhuvan Unhelkar
Sydney, NSW, Australia

Preface

Climate change is a big issue. It has been discussed and continues to be discussed in major forums across the world. The UN Climate Agreement on reducing carbon emissions, which was reached in Paris in December 2015, continues to raise awareness for the need to reduce electricity use through efficiency. Green IT (information technology) provides an excellent way to reduce electricity use and save money. Al Gore's book *Our Choice: A Plan to Solve the Climate Crisis* (Gore, 2009) and other publications continue to point to the urgency of having everyone worldwide help to solve this urgent crisis. Even Pope Francis has weighed in (Pope-Francis, 2015), as have President Barack Obama (Obama, 2017) and former New York City Mayor Michael Bloomberg and former Sierra Club President Carl Pope (Bloomberg & Pope, 2017). Al Gore's 2017 book and movie, *An Inconvenient Sequel: Truth to Power* (Gore, 2017), are both very optimistic on the progress made in fighting climate change.

Our original plan was to write a book on "Using Computer Technologies and Data Science to Make Healthcare Green." That topic was based on my collaboration with Nina Godbole, a PhD student at UPES (University of Petroleum and Energy Studies), in Dehradun, India. Nina was working for IBM India and asked me to be an official mentor for her PhD work on "Green IT and Hospitals." Since then, Nina and I have co-authored several IEEE papers in the area of making healthcare green. As part of her research, Nina has visited hospitals in India and gathered extensive data on green healthcare policies. She met with leaders at several hospitals in India to discuss their views on operational success and social responsibility in green healthcare. Based on our recent research, we decided to expand the scope of the book to further emphasize the role of cloud, green IT, and other areas where we all can contribute to helping save the planet with green healthcare.

Additionally, when I mentioned to my music major niece in 2014 that I was working in the STEM (Science, Technology, Engineering, and Mathematics) area, she said that she and her music major friends wanted that changed to "STEAM" to include the arts. At that time, I thought she was kidding. I now know better. It turns out that green IT is both a great STEM project and a great STEaM (STEM + arts) project. Of course, science and arts have always gone together. The great Renaissance man, Leonardo da Vinci, is famous both as an inventor and scientist and as a great artist. He would certainly be an advocate of STEaM. In our current age of computers and electronic devices, technology and design are very much connected. Steve Jobs, the Apple founder, was more into design than technology. He left the technology details to the "other Steve" (Steve Wozniak). Elon Musk, who launched Tesla electric automobiles, is another modern day Renaissance man who is in the STEaM group. Design and energy efficiency are very much a part of Tesla cars and Musk's SpaceX reusable rockets for launching commercial satellites and even for getting adventurous people to Mars. Musk is also helping create green electric energy through his SolarCity venture. STEaM collaboration is also important for green IT and green healthcare since there is a great need for both design (arts) and technology in making information technology and healthcare green. So, this book title is *Making Healthcare Green: The Role of Cloud, Green IT, and Data Science to Reduce Healthcare Costs and Combat Climate Change*.

This book is about two of the big challenges of our day: the climate crisis and the healthcare crisis. Cost savings are emphasized in addition to the aspect on helping solve the climate crises. The next section of this preface is written by co-author Nina Godbole, which brings in her perspective from India looking at the global healthcare crisis and its connection to the climate crisis. As Nina mentions, this book is written to serve the needs of a wide range of readers. The readers include medical as well as paramedical staff who need a green awareness perspective. The readers also include the IT administrators working in hospitals, the infrastructure personnel in the healthcare industry, and students of business management courses for whom the book should help increase their environmental consciousness, especially while administering and managing hospitals and allied areas. We all need to collaborate in combatting climate change. As our climate change scientist colleagues warn us: "The frog is in the water and the water is heating up."

Healthcare and Green IT: The Duo of Emerging Importance

The nature of changes, now occurring simultaneously in the global environment, with their rate of change as well as magnitude, is like never before in human history and probably in the history of our planet earth.

The healthcare sector is characterized by the fact that it brings together people and institutions involved in providing healthcare to those who need such services in times of stress. Healthcare is not only a complex industry but also a multi-segment industry. It is composed of a number of sub-segments, i.e., hospitals (a major sub-segment), pharmaceuticals, medical insurance agencies, medical equipment manufacturers, dealers and distributors, and diagnostics. The healthcare sector makes heavy use of technology. Technological advances have brought about a revolution in the healthcare industry worldwide – from modern testing techniques to improved surgical equipment, remote health monitoring technologies with the help of modern digital equipment and technology, etc. We now have a number of online healthcare portals. However, the healthcare sector, while being one of the most important sectors, also adds to the carbon footprint. Therefore, the environmental impact of the healthcare sector has also become an important factor globally and continues to draw the attention of government regulators. Thus, the environmental impact of healthcare sector is being felt globally and has come under fire. The energy use of the healthcare sector is growing due to many factors, including the rapid growth and adoption of information and communication technology (ICT) in healthcare. The new IT technologies and applications used in healthcare include "cloud computing" and "mMedicine," i.e., "mobility in Health," eHealth, and tele(health) care for "remote delivery of healthcare services."

A big challenge now facing the healthcare sector is how best to improve the energy efficiency and sustainability of its very complex system. Efforts over the past few years to analyze and create highly efficient data centers present an excellent opportunity for cost-effective green IT at hospitals. Another important dimension is that the costs of healthcare delivery are on the rise, and there is a perpetual need to develop and update facilities. This makes the design and construction of high performance healthcare facilities, a vital priority in the building construction industry, especially in the United States. While green IT and cloud computing constitute one aspect of greening the healthcare sector, high performance green buildings are also important for making healthcare green. It is with this background that we present this book to our readers.

The Message from Our Book

The ultimate goal of this book is to create awareness about making healthcare embrace sustainable practices. A great deal of focused research is required to improve the deployment of IT in the sector, and a lot depends on how healthcare facilities are planned, designed, constructed, and maintained. Toward all that, our book gives a message, and that is, to help SAVE the planet by greening the healthcare sector! This appeal to save our planet is wonderfully presented in J. Krishnamurti's message below:

The death of a tree is beautiful in its ending, unlike man's. A dead tree in the desert, stripped of its bark, polished by the sun and the wind, all its naked branches open to the heavens, is a wondrous sight. A great redwood, many, many hundreds of years old, is cut down in a few minutes to make fences, seats, and build houses or enrich the soil in the garden. The marvelous giant is gone. Man is pushing deeper and deeper into the forests, destroying them for pasture and houses. The wilds are disappearing. There is a valley, whose surrounding hills are perhaps the oldest on earth, where cheetahs, bears and the deer one once saw have entirely disappeared, for man is everywhere. The beauty of the earth is slowly being destroyed and polluted. Cars and tall buildings are appearing in the most unexpected places. When you lose your relationship with nature and the vast heavens, you lose your relationship with man.

Krishnamurti, J. Krishnamurti, Krishnamurti Foundation
Trust Bulletin 56, 1989

Readers for Whom the Book Is Intended

We believe that this book will serve the needs of a wide range of readers – medical as well as paramedical staff from green awareness perspective, the IT administrators working in hospitals, the infrastructure personnel in healthcare industry, and the students of business management courses, to whom the book would help increase their environmental consciousness while administering and managing hospitals and allied areas.

Themes Addressed and the Structure of the Book

Healthcare is a very large area and so is cloud computing and green IT; however, in this book we have tried to connect these three major dots in an attempt to put together a very unique book. Therefore, the themes addressed in this book are: the global importance of the emerging healthcare sector, the role that data science and data analytics play in healthcare, green IT, mobile handheld devices and telemedicine in the context of healthcare, the importance of server virtualization from energy perspective, cloud technologies and carbon footprint of the healthcare sector, data storage strategies for the greening of the healthcare sector, energy reduction in healthcare operations as well as carbon footprint metrics, and last but not the least, the economics of green healthcare. We have also provided case studies to bring in a perspective from the ground-level reality. Given the intended readers of the book, we have divided it into 12 chapters and 3 appendices. Each chapter provides not only summary and conclusions but also a multitude of references for additional reading, making the book also a valuable reference for scholars in the area. The appendices of the book provide ample information that serve as an extended learning of the concepts presented in the book's chapters.

We do hope to see the book well received in the industry as well as among the academicians.

Pune, India   Nina Godbole
New York, NY, USA   John Lamb
May, 2018

Acknowledgments

Many of the ideas and details presented in the following chapters are based on the green healthcare projects and green data center installations we have investigated over the past few years. Therefore, we would like to thank all those involved with the design of these implementations.

Green healthcare requires the collaboration of many groups. So, writing this book also required collaborative input from many different groups and individuals. Dave Anderson, of IBM and a "Green IT Architect," provided the excellent Green IT Checklist in Appendix A. The Global Green and Healthy Hospitals (GGHH) organization provided excellent material for many of the worldwide green healthcare case studies in Chaps. 9, 10, and 11.

Several other organizations provided case studies that are described in Chaps. 9, 10, and 11. The iSixSigma Team provided excellent Six Sigma case study material used in Chap. 9. The Carbon Trust provided case study material in Chap. 11 on identifying carbon reduction opportunities at the University of Reading in the United Kingdom. Jamie Plotnek of the Carbon Trust provided some excellent review comments. The Healthier Hospitals organization and Healthcare Without Harm (HCWH) group provided material for case studies in Chaps. 10 and 11. Co-author Nina Godbole made many visits to three hospitals in India, and her three case studies for green hospitals in India appear in Chap. 11.

One interesting aspect of green IT is the universal interest among diverse groups of people. Almost everyone is interested in green projects and almost everyone uses a PC with connection to the Internet. Therefore, we see the universal appeal of green healthcare and green IT. We would like to thank all of my contacts on this book for their valuable dialog and suggestions. The University of California, San Francisco (UCSF), Office of Sustainability provided excellent case study material used in Chap. 10 in the section "Examples of Greening the Medical Center." Deborah Fleischer and Kailyn Klotz are the authors of the two UCSF sustainability stories included in that Chap. 10 section.

Special thanks go to our colleague in green projects, Dr. Bhuvan Unhelkar of Sydney, Australia. Bhuvan is a thought-leader and a prolific author of 20 books – including several books related to green computing. We were very happy to have a green advocate from Australia join the authors from the United States and India in producing a book on green healthcare. It helped give us the worldwide view we needed on the global scope of green healthcare.

Special thanks also go to our former IBM colleague, Phil Perry, who provided an excellent overall copyedit and peer review of the manuscript based on his long experience with green architecture and green computer technology. Phil's current work position is "software consultant at Catskill Technology Services, LLC," where he is still involved in green computing.

Pace University in Pleasantville, New York, USA, was very supportive in our writing process. Special thanks go to Dr. Jim Stenerson who ran several faculty writing forums where we could exchange ideas with other faculty members on how best to push forward and complete each writing task! Kevin Dake, a chemical engineer from Connecticut, USA, provided excellent peer review comments.

The people at Springer Nature Publishing who worked with us on this project deserve a special thank you for helping complete this project in a relatively short time. The Springer Nature Senior Editor for this book was Christopher Coughlin. Our Springer Nature Project Coordinator, Silembarasan Panneerselvam (Simbu), was always there to help us out in the writing and editing process. We also wish to thank P. Abitha, Project Manager, and her entire team at SPi Technologies India Private Ltd. for smoothly taking our book through the production process.

How to Use This Book

This book is organized largely around the two basic reader groups for a book of this type: (1) healthcare professionals and hospital administrators and (2) information system (IS) professionals, system architects, IT architects, engineers, and other technical groups. If you are more involved in the business aspects of your healthcare system, then you may be most interested in the early chapters on "what to do" in the area of green healthcare. (You may even want to skip some of the more technical "how to" chapters such as Chap. 5 on green IT, cloud technologies, and carbon footprint and appendix A on tools and calculations for estimating healthcare energy and carbon footprint requirements.)

- Chapters 1, 2, and 3 give a background on green healthcare and green IT and should be of interest to all readers. These chapters provide information on the green healthcare challenge and the importance of collaboration across a wide array of technical and regulatory groups.
- Chapters 4, 5, 6, 7, and 8 discuss some of the technical equipment and strategies for saving energy in healthcare. These are the more technical chapters that discuss the "how to" aspects of implementing green healthcare and green IT.
- Chapters 9, 10, and 11, the economics and case study chapters emphasize the "lessons learned" aspect of green healthcare. The case studies include universities and large and small hospitals and span the globe to provide guidance based on country or regional green regulations and issues. These case study chapters should be of interest to all readers.
- Chapter 12 is the summary chapter and also takes a look at the future. This material too should be of interest to all readers.
- The appendices contain checklists for green healthcare, tools for power and cooling estimates, emerging technologies such as grid and cloud computing, background information on the pros and cons of different power generation methods, and information on worldwide electricity average prices for IT. Cloud computing, discussed throughout the book, is a very significant new technology for green healthcare. The recent push for "private cloud computing" will have an impact on all of our green healthcare projects.

This book was written to allow the reader to go directly to a chapter or section of interest and begin reading without having first read all the previous chapters. The intent was to make the content of each chapter as independently intelligible as possible.

Green healthcare is and will continue to be a very interesting and hot topic world-wide. Enjoy this book. It was a pleasure to write.

Nina Godbole and John Lamb
May, 2018

Contents

About the Authors

Nina S. Godbole is an independent consultant and an ex-IBM India employee living in Pune, India. She has written several books on computer technologies. Her books include:

1. *Information Systems Security: Security Management, Metrics, Frameworks and Best Practices*, published by Wiley India, 2008. ISBN-13: 978-8126516926. The second edition of this book is published by Wiley India in 2017. ISBN 978-81-265-6405-7.
2. *Software Quality Assurance: Principles And Practice*, published by Alpha Science Intl Ltd., August 2004. ISBN-13: 978-1842651766. The second edition of this book is published by Narosa in 2017. ISBN 978-81-8487-146-3.
3. *Cyber Security:* Understanding Cyber Crimes, Computer Forensics and Legal Perspectives, published by Wiley India, 2011. ISBN 978-81-265-2179-1.

Nina holds university degrees from IIT Bombay, India. She can be reached at ninagodbole@yahoo.com.

In addition to the books mentioned above, Nina Godbole has published several technical papers. Nina was a PhD student at UPES (University of Petroleum and Energy Studies), in Dehradun, India. Nina worked for IBM India for over 15 years and requested that John Lamb work with her as an official mentor for her PhD work on "Assessment of Benefits through Private Cloud adoption of Electronic Health Records (EHR) in Indian Hospitals." As part of her research, Nina embarked upon an extensive study of literature published in the related area and also met with relevant authorities at some leading hospitals in India to gather data about the views of those hospitals on their awareness about "green" being an important aspect of hospitals' operational success as well as a way to demonstrate their social responsibility in view of the climate-related challenges that face us. Nina's discussion with the concerned authorities and management members at the hospitals included understanding their perspectives about Green IT practices and the use of related techniques at the hospitals to form green healthcare policies. She can be contacted at ninagodbole@yahoo.com

John P. Lamb retired from IBM on December 31, 2013, after more than 40 years with the company. He is currently an adjunct professor of mathematics at Pace University in Pleasantville, NY. A senior member of the IEEE and ASME engineering societies, he has published more than 60 technical papers and articles. He has also authored five books on computer technologies including the May 2009 book: *The Greening of IT: How Companies Can Make a Difference for the Environment*, ISBN 0137150830.

John holds a PhD in engineering science from the University of California at Berkeley and a BA in mathematics from the University of Notre Dame. He can be reached at jlamb@pace.edu.

John Lamb has previously authored five technical books, three for McGraw-Hill and two for Pearson Education/IBM Press:

1. *Lotus Notes Network Design*, published by McGraw-Hill, 1996. ISBN 0070361606.
2. *Lotus Notes and Domino Network Design*, published by McGraw-Hill, 1997. ISBN 0079132413.
3. *Lotus Notes and Domino 5 Scalable Network Design*, published by McGraw-Hill, 1999. ISBN 007913792X.
4. *IBM WebSphere and Lotus: Implementing Collaborative Solutions*, published by Pearson Education/IBM Press, September 2004. ISBN 0131443305.
5. *The Greening of IT: How Companies Can Make a Difference for the Environment*, published by Pearson Education/IBM Press, May 2009. ISBN 013715083.

In addition to the books and papers listed above, John Lamb continues to publish technical papers and articles some of which are with Nina Godbole.

Chapter 1
Healthcare: An Emerging Domain and Its Global Importance

We are using resources as if we had two planets, not one. There can be no "plan B" because there is no "planet B."

— Ban Ki Moon – former UN Secretary General

As more and more people understand what's at stake, they become a part of the solution, and share both in the challenges and opportunities presented by the climate crises.

— Al Gore – former USA Vice President and Nobel Peace Prize Winner – on Global Warming

As discussed in the preface, the climate crisis and the healthcare crisis are two of the biggest issues of our age. Solutions to the two crises are related. Green healthcare provides an excellent way to reduce electricity use and reduce carbon emissions and, at the same time, save money. Using technology to reduce costs and improve service is crucial to helping solve the healthcare crisis. The climate crisis and healthcare crisis are worldwide problems, and many groups and publications continue to point to the urgency of having everyone worldwide help to solve these urgent crises. Even Pope Francis has weighed in (Pope-Francis, 2015), as has President Barack Obama (Obama, 2017). This first chapter shares background information on healthcare and underlines the urgent need for improvement on a global basis.

The Healthcare Domain in the Global Context

Healthcare is not just another service industry; the healthcare sector is characterized by the fact that it brings together people and institutions involved in providing healthcare to those who use such services in times of need and stress. There is a vital link between citizens' health and a nation's economy. It is not a surprise then that the issues of the healthcare industry have moved on the top page of the OECD (the Organization for Economic Co-operation and Development) agenda (Frenk Julio, 2004).

© Springer International Publishing AG, part of Springer Nature 2018
N. S. Godbole, J. P. Lamb, *Making Healthcare Green*,
https://doi.org/10.1007/978-3-319-79069-5_1

The healthcare sector makes heavy use of technology. Technological advances have brought about a revolution in the healthcare industry worldwide – from modern testing techniques, to improved surgical equipment and remote health monitoring technologies. In addition, the healthcare industry plays a vital role in the economy of our nations. For example, the healthcare industry dominates the GDP (gross domestic product) of any country. It also determines employment capital investment export status, etc. The healthcare segment provides employment opportunities to many individuals directly or indirectly associated with the healthcare sector or other associated sectors, related to the healthcare industry in some way or the other. As an example, consider this; a pharmaceutical company operates in more than 45 countries and employs 2500 people in India (Mahmud Adeb, The Role of The Healthcare Sector in Expanding Economic Opportunity, 2007).

With a few exceptions, in the past 50 years, there has been a tremendous growth in the global economy, although not all sectors of business have necessarily contributed to it. If we are to continue keeping our economic growth and retain its progress, our vitality in terms of "health" is no doubt very crucial – this is where the healthcare domain is of utmost importance. In terms of improving the quality of healthcare services delivered to citizens and in terms of extending and/or improving the access to medical care, global multinational healthcare companies have a pivotal role to play. While the major focus of healthcare industry is centered at (i) improving the quality of healthcare services provided, (ii) making healthcare services affordable, i.e., the reducing costs, and (iii) extending the reach of healthcare services by improving access to healthcare, the healthcare sector also has a participation in expanding global economic opportunities.

As explained in the next section, healthcare is a very broad industry, consisting of multiple sectors (refer to Fig. 1.1). Among the sub-sectors of the healthcare

Fig. 1.1 Healthcare industry breakdown

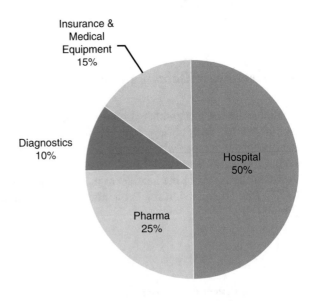

industry, hospitals and pharmaceuticals form a large one. For the purpose of this book, we focus on the largest sector of the healthcare industry, i.e., hospitals. There are a number of other players in the healthcare sector (apart from hospital chains and pharmaceutical companies). Most notable among them are the healthcare insurance providers – they have their own "value chain" and employ a large number of staff.

A research study, (Doeksen and Schott, 2003) carried out in the US state of Oklahoma, about rural and remote health (which relates to factor (iii) mentioned earlier, i.e., extending the "reach" of healthcare services) demonstrated the economic importance of rural healthcare in terms of its quantifiable impact on employment and payroll. Studies in healthcare in other developing economies, such as the People's Republic of China (PRC), also show the strong correlation between the healthcare sector and economic growth (Brant Simone, 2006; Brian, 2003; Chelala Cesar). Regarding factor (i) related to cost of healthcare services, one of the driving pressures on the healthcare industry is the need to deliver healthcare to more patients for less money. This means that the relevant stakeholders of the industry need to continually strive to identify the waste in the healthcare delivery system, i.e., the "ineffective" elements that add to raising the cost of healthcare service delivery.

Therefore, one of the challenges would be to analyze cost structures to make an efficient transition from "treatment-based medicine" to "outcome-oriented healthcare delivery." This points to the need for the healthcare industry to invest more and more in the adoption of digital technology to reduce reliance on face-to-face care. The result of effective deployment of this strategy would be twofold (1) providing more care at home or closer to home and (2) extending the access to healthcare systems by enabling patients toward a self-help model.

The Healthcare Industry: An Overview

Let us understand the structure of the healthcare sector. Healthcare is a complex and heterogeneous industry consisting of multiple sectors. According to some industry experts, it is a *fragmented industry* (BusinessDictionary.com, n.d.). A *fragmented industry* means an industry in which no single enterprise has a large enough share of the market to be able to influence the industry's direction. In other words, it is a marketplace where there is no single company that can exert enough influence to move the industry in a particular direction. Such a market consists of several small- to medium-sized companies that compete with each other and large enterprises. From a domain perspective, the healthcare area consists mainly of (1) pharmaceuticals, (2) medical areas, and (3) health service organizations/healthcare organizations. Hospitals form a large part of the healthcare industry (see Fig. 1.1). Globally "healthcare" is emerging as an important domain in terms of market size.

The Healthcare Industry: The Challenges

In an earlier section, we mentioned the three major challenges faced globally by the healthcare industry – (i) improving the quality of healthcare services provided, (ii) making healthcare services affordable, i.e., the reduction of costs, and (iii) extending the reach of healthcare services by improving access to healthcare. Let us understand these three challenge factors.

The healthcare industry worldwide is grappling with the challenge of how to reduce the cost, how to improve the quality of service, and how to extend the reach (Deloitte, 2012) (OECD – Organisation for Economic Co-operation and Development, 2013). In the United States, healthcare a very competitive sector, is facing the same problem (Porter, 2007). Strategically, as a business sector, healthcare needs to focus on creating customer value above all (Judge & Ryman, 2001). The cost dimension is important from the healthcare value-chain perspective (Shoui, 2013). The challenge of making affordable healthcare delivery is noted in one more reference (Govindarajan & Ramamurti, 2013).

The four-faced challenge for the healthcare sector is explained in the work just mentioned; although this article appeared many years back, it is pertinent even today as this work reviews the changes that have taken place in the healthcare domain during the four decades since the end of World War II. The challenges mentioned in other work are: (1) quality of (health)care, (2) cost containment, (3) technological changes, and (4) insurance related to healthcare (Weisbrod, 1991).

A number of strategies are proposed to control the rising cost of healthcare. Most of such strategies are aimed at reducing medical resource consumption rates. It is argued that these approaches may be limited in effectiveness due to the relatively low variable cost of medical care. Variable costs (for medication and supplies) are saved if a facility does not provide a service, while fixed costs (for salaried labor, buildings, and equipment) are not saved over the short term when a healthcare facility reduces service. Cost analysis (in the United States) was done based at a hospital having nearly 114,000 emergency department visits, 40,000 hospital admissions, 240,000 inpatient days, and more than 500,000 outpatient clinic visits. It was found that the majority of costs in providing hospital service were related to equipment, buildings, salaried labor, and overhead, which are fixed over the short term. The high fixed costs emphasized the importance of adjusting fixed costs to patient consumption to maintain efficiency (Roberts et al., 1999).

Another study shows that implementing health management information systems will translate directly to efficiency gains (Noir & Walsham). The Indian healthcare sector, too, faces many of these challenges. Thus we can say that while delivery of affordable and quality healthcare to billions of people worldwide remains a huge challenge, it also presents huge opportunities for reaping the benefits of information technology (IT) in hospitals which constitute the largest chunk of Indian healthcare sector too. India needs 3 million more hospital beds to match the global average of 3 beds per 1000 population (Dinodia Capital Advisors, 2012; Gyan Research and Analytics, 2012). India's vision for health informatics is described in a report by the

Center for Development of Advanced Computing (Ramakrishnan, 2008). The healthcare field has limited literature published on the subject of global deployment, diffusion, adoption, use, and impact of information communication technologies in the context of health informatics and electronic healthcare applications. As a result, healthcare professionals spend a great deal of time searching and gathering information from various sources (Khalil Khoumbati, 2010).

A key issue now is reducing healthcare delivery costs, increasing the quality as well as the operational efficiency of healthcare operations, and improving the "reach" of healthcare delivery. Given the importance of the healthcare sector, its challenges, and the fact that hospitals constitute a major portion, it is worthwhile examining healthcare sector from these three perspectives, i.e., (1) quality, (2) cost, and (3) reach. Transformation of healthcare requires high-quality data for ensuring that it is truly patient-centered and clinically led. Collecting and sharing accurate information with health services providers, patients, and the public will help to assess safety issues and identify areas where outcomes and patient experience in healthcare organizations can be improved. Worldwide, for healthcare, there are significant benefits of having detailed information available for each hospital. As citizens, we should be in a position to compare the quality of healthcare provided by different hospitals, different hospital teams and wards, and individual clinicians (Bishop, 2014). A study was undertaken on the changing trends in hospital information systems, keeping in focus "performance parameters" for hospital as well as regulatory requirements such as HIPAA. This study was qualitative in nature and was based on existing survey data and specific successful case studies (Balaraman & Kosalram, 2013). From a regulatory perspective, the "triple" challenge faced by the healthcare domain is a topic of significant interest (Godbole & Lamb, The triple challenge for the healthcare industry: Sustainability, privacy, and cloud-centric regulatory compliance, 2013).

Green practice has been considered as a way for innovation (Elzen, Geels, & Green, 2004). "Green IT" ("green computing") involves using computer resources in an efficient way and it is a relatively new concept. Green IT or green computing aims to reduce the carbon footprint while allowing to save money. The Gartner definition for green IT is "optimal use of information and communication technology (ICT) for managing the environmental sustainability of enterprise operations and the supply chain, as well as that of its products, services and resources, throughout their life cycles" (Mingay, 2007). CIO attitude worldwide is changing about green IT (Gedda, 2011). Green IT is considered to be "better IT" (Varon, 2007).

Healthcare and Green IT

Green IT initiatives ought to begin with manufacturers committed to produce environmentally friendly products and encouraging IT departments to consider more environmentally friendly options like virtualization, power management, and proper recycling habits. The number of servers reduced/rationalized through virtualization can reduce power consumption in a data center by 200–400 W per server. This is the equivalent of about $380 per year, per server, factoring in the energy costs of air conditioning to cool the unit (Sullivan, 2009). Some recommendations are using low-emission building materials, recycling, using alternative energy technologies, and other green technologies. A holistic approach to green IT is recommended by industry experts (Harris, 2008). In fact, one of the reasons toward green practices in business is for attracting investors (Gorbett, Salvaterra, & Skiba, 2005).

The green IT life cycle includes key aspects such as green data centers, green cloud computing, green data storage, etc. (Murugesan & Gangadharan, 2012). According to one article, the boom in cloud-centric data-center construction is driven by healthcare technology (PR Newwire, 2012). However, there is a debate as to how "green" is cloud computing (Swan, 2011). In yet another piece of literature, it is argued (by providing four reasons) why cloud computing is a green solution (Mines, 2011): (1) resource virtualization, enabling energy and resource efficiencies; (2) automation software, maximizing consolidation and utilization to drive efficiencies; (3) usage-based payment (pay per use), self-service, encouraging more efficient behavior and life-cycle management; and (4) multi-tenancy, delivering efficiencies of scale to benefit many organizations or business units. The mention of cloud computing as one of the examples of "green IT" is noted in one literature (Beom).

According to another report, cloud computing in healthcare was estimated to reach $5.4 billion by 2017 (Horowitz, 2012). The business perspective on cloud computing is noted in another paper (Marston, Li, Bandyopadhyay, Zhang, & Ghalsasi, 2011). In this article, cloud computing issues from the stakeholder perspective are addressed. Cloud computing technology's potential to significantly reduce the energy consumption and carbon footprint of organizations is mentioned in one more piece of literature reviewed. From a green perspective, cloud computing is claimed to be green according to one article reviewed. Cloud computing is discussed in yet another article (O'Donnell, 2014) arguing that usually cloud computing is more efficient than enterprise computing because a cloud typically leverages shared infrastructure and is high on average utilization of resources. Two more experts support the premise that cloud computing is very energy efficient if we consider new, leading edge cloud implementations and if we keep aside the fact that ease of access probably means greater use and that the data centers themselves could be more efficient.

It is argued that cloud computing is more energy efficient than traditional computing, and therefore it can be said that cloud computing is green. Cloud computing technology makes use of better solutions to reduce power consumption by data centers. For improving the energy efficiency of data centers, cloud computing technology provides many solutions such as network power management, chip multiprocessing energy efficiency, data-center power capping, storage power management solution, etc. However, among all these approaches, virtual machine (VM) technology has emerged as best for deployment. Virtual machine technology (such as Xen, VMWare, Microsoft Virtual Servers, the new Microsoft Hyper-V technology, etc.) enables multiple operating system environments to coexist on the same physical computer, in strong isolation from each other. VMs share the conventional hardware in a secure manner with excellent resource management capacity, while each virtual machine is hosting its own operating system and applications.

The virtual machine platform facilitates server consolidation and colocated hosting facilities. Virtual machine migration, which is used to transfer a virtual machine across physical computers, has served as the main approach to achieve better energy efficiency of data centers. Using virtual machine and virtual machine migration technology helps to efficiently manage workload consolidation and therefore improves the total data-center power efficiency. However, there is further debate given the opposite view of other IT experts who argue that a cloud is where the machinery is (i.e., it is not in an organization's data center per se), while virtualization allows us to provision a big server into individual virtual machines. Therefore, virtualization, on a per-virtual-server basis, may save energy if an organization/enterprise does not expand the number of servers they deploy. Further, it is argued that there is a wrong notion that business will not use more servers and therefore have a smaller physical footprint. In fact, virtualization might provide the stimulus to put up more applications causing an actual increase in energy consumption because the cost of a VM (Virtual Machine) is less expensive and faster than a physical server. In their point against cloud computing being green, it is further mentioned that today's blade centers (used by most IT departments to deploy virtualized servers) are considerably more power hungry. They argue that the most popular blade center consumes more power (including HVAC) than an old Sun V990 of equivalent computing power. So, the contention is that this increased power problem is a temporary aberration in order to accommodate the virtualization-hungry IT departments, which are being pressured by healthcare IT, and especially hospital IT, to cut costs. To sum up, we can relate this in terms of green IT benefits to the healthcare sector (whose largest sub-segment is hospitals) by noting the fact that the data center is a hot market for the healthcare sector.

Opportunities for cloud computing in healthcare (Canada Health Infoway Inc, 2012) are noted in one more literature reviewed. As per an announcement made by IBM, many healthcare clients are adopting EHR (electronic health records) with cloud-based services. A major rationale for this, as mentioned in the report, is that many healthcare providers struggle to manage the high costs of providing quality patient care.

The low EHR utilization rate can be ascribed to providers having challenges with investment of time and resources into information technology (IT) for improving their operations. This makes acquisition of cloud-based service technology compelling in the healthcare industry (Yanos, White, Suprina, Parker, & Sutton, 2009). The advantages as well as the downside of cloud-hosted EMR (electronic medical records) are noted in the literature. The advantages include standardization of infrastructure for healthcare IT solutions, faster IT resource provisioning (hardware), higher scalability, and greater sharing of patient health information. Security is noted as one of the main concerns with the use of cloud computing in healthcare (Jaswanth, Durga, & Kmar, 2013). Trustworthy Cloud for Health Platform has been mentioned. In this work, an approach to build a trustworthy healthcare platform cloud, based on a trustworthy cloud infrastructure, is proposed. It is claimed that such an approach is compliant with EU data protection legislation. They have described results from the recent EU TClouds project as a possible solution toward trustworthy cloud architecture, based on a federated cloud of clouds.

It is further claimed this helps enforcing security, resilience, and data protection in various cloud layers for provisioning trustworthy IaaS, PaaS, and SaaS healthcare services (Deng, Nalin, Petković, Baroni, & Marco, 2012). There is a difference between virtualization and cloud computing (Intel). Cloud computing is possible without virtualization. In a cloud environment, energy consumption of underutilized resources accounts for a substantial amount of the actual energy use. A better energy efficiency resource allocation strategy takes into account resource, which, in clouds, extends further with virtualization technologies (Liu et al., 2012). Basically, green IT is classified into two types: (1) of IT, i.e., energy saving by reduction of IT equipment, and (2) by IT, i.e., society's energy saving by using IT (e.g., to reduce travel).

There are benefits from adopting green IT (Greenpeace International, 2010; Kim, N.A.). In this study, it is assumed that cloud computing, which is a one of the IT infrastructure provisioning methods, is green. This assumption has support in a work wherein the CLEER model (Cloud Energy and Emission Research Model) has been proposed. It is an open access model for accessing the net energy and emission implications of cloud services at different levels. According to this model, moving applications such as email, spreadsheets, CRM software, etc. to the cloud has the potential to save a large amount of energy. In a study conducted in the United States, business software applications were selected in the analysis of the CLEER model. The study concluded that in spite of many difficulties in using green computing for establishing energy-efficient systems, there is a positive impact on the environment (AGRAWAL, 2013).

Another study claims that moving applications to the cloud can save 30% or more in carbon emissions per user (Microsoft and Accenture, 2010).The study focused on hospitals – as mentioned earlier, hospitals constitute the largest proportion of the healthcare sector (refer to earlier Fig. 1.1). ICT (information and communication technology) systems typically account for about 25% of direct electricity use in commercial office buildings and in buildings or locations with a high density of IT gear. ICT represents 2–2.5% of the total global carbon emissions, equivalent to the global aviation industry.

Green Healthcare: Its Importance for our Planet

As the world becomes environmentally conscious, healthcare gets compared with other industries in terms of the carbon footprint generation (Energy Information Administration (EIA), 2006; Singer & Tschudi, 2009; Singer, Coughlin, & Mathew, 2009). Greening of the healthcare sector is important from the carbon footprint perspective. For example, in England, the National Health Service (NHS) has calculated its carbon footprint at more than 18 million tons of CO_2 each year – 25% of the total public sector emissions. It is said that, in spite of worldwide interest in healthcare sustainability and given that healthcare is a very large industry in the United States, the carbon footprint of the US healthcare industry has not been estimated (JAMA Editor, 2009). Moreover, healthcare is the second most energy-intensive commercial building sector. Hospitals have a mission-related incentive to reduce carbon burden. Hospitals save millions of dollars through climate mitigation (Leetz, 2011). Greening of healthcare as an area of focus has been given a greater attention over the years (Godbole, 2011).

Green computing contributes toward energy efficiency. According to the green computing energy efficiency guide, "green computing" is no longer just a buzz word nor is it a new fad (TechTarget, Green computing energy efficiency guide, 2014). A hospital's mission is related to the issue of sustainability. After all, without a sustainable operation, there is no hospital. Usually, sustainability goals refer to energy conserved, money saved, waste diverted, water recycled, or some other relevant metric. However, a clear nexus between sustainability and a hospital's mission, in a language that is easy to understand and with an emphasis on facts, does not occur so easily. The relation between sustainability and "green mission" must extend to effective management of the healthcare environment, social interaction between patient and healthcare provider, community-based healthcare approaches, and the utilization of current technology (Siebenaller, 2012).

As mentioned earlier, in an environmentally conscious world, healthcare gets compared with other industries in terms of the carbon footprint generation. EPA-based calculations reveal that emissions from energy production ultimately consumed by healthcare facilities – including sulfur dioxide, nitrogen oxide, carbon dioxide, and mercury – cause an increased disease burden in the general public, including conditions such as cardiovascular diseases, asthma, and other respiratory illness. In countries around the world, the health sector plays a significant role in the economy. The sector purchases everything from linens to computers, medical supplies, and transportation vehicles and does so in large volume. The NHS in England calculates that it spends 20 billion pounds a year on goods and services, which translates into a carbon footprint of 11 million tons – 60% of the NHS's total carbon footprint. The health sector can reap its economic leverage by shopping green. This means purchasing environmentally friendly and sustainable materials and products whenever possible, including products with minimal carbon impact.

Transportation is a major source of greenhouse gas emissions, and it is a global issue as far as the carbon footprint is concerned. The health sector – with its fleets of hospital vehicles, delivery vehicles, and staff and patient travel, is a transportation-intensive industry. In England, for example, transportation is responsible for 18% of the NHS's (National Health Service) total carbon footprint (World Health Organization). To achieve energy and carbon savings in hospitals, chief executive needs to demonstrate a strong leadership and as a focal point for showing the commitment to make hospitals "green." "Energy champions" are required at the board level of hospitals (Carbon Trust, 2010). Energy conservation in hospitals is important, not only for reducing carbon footprints but also as a way to increase profitability (Sullivan, 2012). In the United States, healthcare ranks as the country's second most energy-intensive industry, with hospitals spending more than $10 billion each year.

Hospitals are the sector's largest energy consumer and producer of greenhouse gases (GHG). The healthcare industry's reliance on conventional, nonrenewable energy sources (oil, coal) contributes disproportionately to the emission of GHG, driving climate change and impacting public health from air pollution (Premier Inc, 2013). According to a survey based on responses from 1056 healthcare compliance professionals, the compliance dimension for healthcare industry is also large. For example, HIPAA/HITECH is the major one (HIPAA is the Health Insurance Portability and Accountability Act, and HITECH is the Health Information Technology for Economic and Clinical Health Act) (Becker, 2013). Interestingly, in one study, it was found that HIPAA authorization forms are written at too high a level for most of the population (Collins, Novotny, & Light, 2006). Through a green ICT strategy, healthcare organizations see fewer bills, generate lower levels of carbon emissions, and experience an easier transition into mandated practices. However, only a limited amount of work has been done to identify the ICT application in redesigning healthcare management system as a whole. ICT (information and communication technology) has the potential to improve the environmental footprint of the healthcare industry.

Business process reengineering (BPR) based on ICT (information and communication technology) helps hospitals achieve their "go green" mission. Toward this, there is a business process engineering-based study that focuses on the need for ICT practices that can be implemented in the healthcare industry to transform hospital practices greener. It is possible to adopt a model for implementing ICT techniques in the healthcare industry in order to make the hospitals go green (Hussain & Subramoniam, 2012).

According to a Department of Energy study, the healthcare industry plays a critical role in climate change mitigation. Medical facilities make a highly intensive use of energy. It turns out that hospitals use about two times as much total energy per square foot as traditional office space. We are facing a real issue. The rise in the Earth's temperature will continue, impacting the Earth's physical and biological systems and our public and environmental health as long as we do not reduce the emissions of these greenhouse gases (GHGs). Over many years, our planet has witnessed extreme weather events, which, in turn, result in infectious disease, negative impact on our fresh water supplies, and health-related issues owing to the rising levels of air pollution. Mitigating climate change is the challenge where the healthcare industry alone cannot be sufficient to measure up to it. The challenge will require collaborative action at multiple levels.

There are surveys (e.g., (Rebecca, 2015)) that expose the appalling condition of healthcare facilities in terms of the building used. There are healthcare facilities that run in aging, energy-inefficient buildings. The "green washing" in the healthcare industry is reported in a survey (Steve, 2012). There are many services that operate around the clock, 365 days a year. According to Practice Greenhealth (Practice Greenhealth), healthcare spends $5.3 billion on energy every year. Medical facilities also consume large amounts of other resources. For example, it was found that in the San Francisco Bay Area, a medical center generated an average of 6 tons of solid waste every day. In the United States, health represents 1/7th of the country's economy, and given that construction in the healthcare industry adds over 100 million square feet of medical building space every year, reducing health care's greenhouse gas emissions is indeed a task with immense importance. The healthcare sector can demonstrate its commitment to environmental cause through the development and implementation of climate change action plans.

The healthcare industry can consider a number of measures toward honoring their commitment to making the sector green: these measures can come from areas such as (1) transportation associated with the sector, (2) energy operations, (3) "greening" the building facilities used by the healthcare sector, (4) effective management of "waste" in the health sector, and (5) managing ancillary services (e.g., the food-related supplies) in the hospitals. In the next section, green ideas for each of the mentioned areas are presented.

The Greening of Health from a Transportation Point of View

It is an irony that while rising emission levels worldwide are known to result in health issues, the healthcare sector itself is a significant source of carbon emissions around the world and therefore a contributor to climate change trends that undermine public health. The American healthcare sector accounts for nearly a tenth of the country's carbon dioxide emissions, according to a first-of-its-kind calculation of healthcare's carbon footprint (Health care accounts for eight percent of US carbon footprint, 2009). The healthcare industry makes its "contribution" to our planet's carbon footprint and thereby to climate change, the way it deploys technologies, through the energy and resources it consumes and the way it operates its (nongreen) buildings. In more concrete terms, all of the activities listed below add 1 kg of CO_2 to the carbon footprint:

1. Traveling by public transportation (train or bus) a distance of 10–12 km
2. Driving a car a distance of 6 km (under the assumption of 7.3 L of petrol per 100 km)
3. Flying with a plane a distance of 2.2 km
4. Operating a computer for 32 h (assuming 60 W of electricity consumption)
5. Producing five plastic bags/two plastic bottles/one-third of an American cheeseburger (the production of each cheeseburger emits 3.1 kg of CO2!)

In the United States, transportation accounts for 27% of the United State's greenhouse gas emissions; that makes it the second largest source of emissions for the country. Transportation is also the most rapidly growing emission sector. The process of powering motor vehicles produces large amounts of CO_2 and traces of methane (CH_4) and nitrous oxide (N_2O) and potent greenhouse gases. Healthcare facilities heavily depend on transportation systems for the physical movement of patients, workers, supplies, and waste. Therefore, careful analysis and redesign of hospital transportation can have climate change mitigation benefits. In this regard, the effective deployment of "remote care," mMedicine (mobile medicine) are significant in addition to helping extend the reach of and access to healthcare; because that way transportation sector emissions could be reduced by reducing total travel or making travel less emission intensive. Reducing transportation GHG (Green House Gas) emissions can also lower other vehicle emissions with known health impacts:

1. Nitrogen oxides (a precursor to smog), benzene (a carcinogen)
2. Particulate matter (a trigger for respiratory illness and symptoms)
3. Volatile organic compounds (some of which are potentially hazardous and a precursor to smog)
4. Carbon monoxide (an acute toxin)

The Greening of Health from the Energy Operations Point of View

In the healthcare sector, hospitals that form a major segment (refer to Fig. 1.1) consume a large amount of energy owing to the peculiarity of its operations. The energy use of hospitals has been growing due to many factors. These factors include the rapid growth and adoption of information and communication technology (ICT) in healthcare. The new IT technologies and applications used in healthcare include "cloud computing" and "mMedicine," i.e., "mobility in health," eHealth, and tele(health) care for remote delivery of healthcare services. In general, the healthcare industry needs to reap the benefits of emerging technologies such as mobile computing and cloud computing, along with the use of health information technology (HIT) to help solve the ever-growing operating cost problems. One challenge facing the healthcare sector is how best to calculate the energy efficiency of this very complex sector. The work done over the past few years to analyze and create very energy-efficient data centers presents an excellent opportunity for cost-effective green IT at hospitals (Godbole & Lamb, 2014).

Energy used in hospitals' facility operations is a significant source of greenhouse gas emissions. Fossil fuels are burnt to heat, cool, and power buildings. In the process of burning those fossil fuels, large amounts of not only CO_2 but also traces of CH_4, N_2O, and sulfur hexafluoride (SF_6) are produced, and these are the most potent GHGs, i.e., greenhouse gases.

The percentages that account for energy use are – water heating accounts for 28% of the energy used in healthcare buildings, space heating accounts for 23%, lighting 16%, and office equipment 6%. In addition to these, there are other energy needs such as refrigeration, air conditioning, and cooking, and together they account for the remaining 27%. Healthcare facilities can simultaneously reduce greenhouse gas emissions (GHG) and save money by conserving energy, purchasing green products (products with minimal supply chain), making operations more energy efficient, and purchasing renewable energy or installing renewable energy infrastructure. For example, LED (light-emitting diode) lighting products produce light approximately 90% more efficiently than incandescent light bulbs. An added benefit to reducing energy is that the reduction in energy use also improves outdoor air quality, thereby benefiting human health by lowering particulate and toxic chemical emissions produced due to fossil fuel combustion and electric power generation.

Green Building for the Greening of Health

Although this aspect does not constitute "green IT", important as one of the factors that contributes to making healthcare "green." Sufficient reduction in energy and greenhouse gas emissions (GHG) can be achieved by incorporating sustainability into the design and construction of hospitals built and landscaping environment (CII – Confederation of Indian Industry, 2010; Gail, 2002; Healthcare without Harm, 2002). Building designers worldwide strive to incorporate design features that can reduce the energy required to operate healthcare facilities. Extraction of raw materials used for building construction cause ecological disruption, threatening natural carbon sinks. The need of the hour is to select environmentally friendly construction material selection (e.g., mud bricks, bamboo, adobe bricks, rammed earth constructions etc. However, these are unconventional materials that may not be very suitable for commercially used buildings). The choice of furnishing can also help mitigate climate change by protecting this natural sequestration, as well as reducing fossil fuel energy required for material manufacture and transport.

Finally, all the factors mentioned earlier can be put in an action plan – this aspect is addressed in Chap. 8 (The Need for Standard Healthcare and Hospital Energy Use and Carbon Footprint Metrics).

Chapter Summary and Conclusions

In this chapter, we explained the importance of the healthcare industry along with some of its significant features. We provided a broad overview of the healthcare domain, explaining its segmentation, key players, etc. We then addressed the key challenges faced by the sector. The chapter ends with the relation of the healthcare sector to climate change which is one of the toughest challenges of this century for our planet.

Additional Reading

A Guide to Global Warming Theory. Published in the Heritage Foundation and available at http://thf_media.s3.amazonaws.com/1992/pdf/bg896.pdf.

Overcoming Fragmentation in Health Care. https://hbr.org/2013/10/creating-a-sustainable-model-for-health-care.

2016 GLOBAL HEALTH CARE OUTLOOK. (November 2015). Published by the Carlyte Group, Accessed at https://www.carlyle.com/sites/default/files/market-commentary/october_2015_-_global_health_care_investment_outlook.pdf. August 21, 2016.

The Indian Model for Low Cost Delivery of Healthcare Delivery. Prepared for Department for International Development (DFID). (December 2014) Accessed September 25, 2016. http://www.psp4h.com/wp-content/uploads/2014/05/Understanding-the-India-Low-Cost-Model-of-Healthcare-Delivery-3.pdf.

CEWIT 2013, (October 21 & 22). The 10th International Conference and Expo on Emerging Technologies for a Smarter World Proceedings. Accessed September 25, 2016 http://www.cewit.org/conference2014/pastconference/conference2013/CEWIT_13_Proceedings.pdf.

Protecting Public Health from Climate: a call for global action. Accessed 12 Sept 2016 http://www.wma.net/en/20activities/30publichealth/30healthenvironment/Durban_Global_Climate_and_Health_Call_to_Action_Final.pdf.

The Greening of Health: the convergence of health and sustainability. Accessed 5 Nov 2016 http://www.iftf.org/uploads/media/SR-1215%20green%20health%20 3.27jr_sm.pdf.

Green Guide for Healthcare. Accessed 5 Nov 2016 http://www.gghc.org/.

Greening UW Health Sciences. Vogt, J., Power, J., Bucy, J. Accessed 12 Nov 2016 at https://green.uw.edu/sites/default/files/docs/capstonereport_greening-health-sciences.pdf.

Chapter 2
The Role of Data Science and Data Analytics in Healthcare

Based on a survey of over 4000 information technology (IT) professionals from 93 countries and 25 industries, business analytics was identified as one of the four major technology trends in the 2010s.

— IBM Tech Trends Report (2011)

One of the complexity dimensions for healthcare is the massive data it generates. In 2012, the estimated *size* of the data generated by worldwide digital healthcare was 500 petabytes (a petabyte is 10^{15} bytes or 1 million gigabytes). The growth of healthcare data size is projected to continue exponentially. By 2020, healthcare data is expected to reach the size of 25,000 petabytes (Sun & Reddy, 2013). Thus, healthcare sector indeed has the "Big Data" challenge. The ever-rising growth of healthcare data will serve no purpose unless it helps the sector. The essential question is "Can Big Data and Data Analytics used by healthcare help it learn from the past to become 'smart' in the future?" There is already a thought on moving from "treatment-based" practices to "outcome-based" practices. If people do not get the cure in spite of the large expenditures on IT in healthcare, then it is a losing battle for the sector.

Data Science and Big Data

The field of "data science" has grown phenomenally in the past few years. This is because over the past 10 years or so, there has been a tremendous increase in the amount of data around us – both the new data generated and the data retained by organizations; mainly for statutory and analytical purposes. Figure 2.1 depicts at a broad level how Big Data analytics is used in healthcare.

According to the IDC definition, *Big Data* is about a new generation of technologies and architectures designed to extract value economically from very large volumes of a wide variety of data by enabling high-velocity capture, discovery, and/or

© Springer International Publishing AG, part of Springer Nature 2018
N. S. Godbole, J. P. Lamb, *Making Healthcare Green*,
https://doi.org/10.1007/978-3-319-79069-5_2

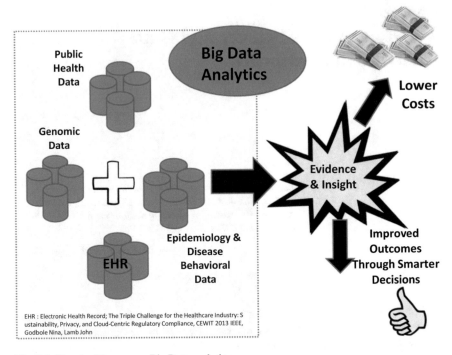

Fig. 2.1 How healthcare uses Big Data analytics

analysis. *Big Data* are high-volume, high-velocity, and/or high-variety information assets that make it mandatory to use new forms of processing and technologies to facilitate improved decision-making, insight discovery, and process optimization. *Data science* deals with the mining of knowledge from large volumes of data that are both structured and unstructured. In this sense, "data science" can be considered as a continuation of the field of data mining and predictive analytics, also known as knowledge discovery in databases (KDD). "Unstructured data" can include videos, social media, emails, photos, and other contents generated by users. *Big Data* are multidisciplinary in nature and that presents many challenges in their handling. While "Big Data" is *not* new, the tools used to handle it are.

The Relation Between Cloud and Big Data

There is a symbiotic relationship between cloud computing and Big Data – *cloud computing* resources are needed to support the storage projects that involved analysis of *Big Data*, and moving to cloud-based computing becomes a business case for Big Data. Data scientists are required for analysis of Big Data. Thus, data science is the analytical "glue" for Big Data and its underlying cloud environment.

An Overview of "Analytics" and Healthcare Data Complexity

The main sources and techniques for Big Data in healthcare are (1) structured EHR data (electronic health records), (2) clinical notes that are unstructured, (3) medical imaging data, (4) genetic data, and (5) other data sources, for example, epidemiology and data on disease behavior (refer to Fig. 2.1). Given the complexity of healthcare data (the fact that it involves "Big Data"), the challenges in analytics include but are not limited to data search, data capture, data storage, data sharing, and data analysis. Another dimension to the complexity of healthcare data analytics comes due to the adoption of electronic health record (EHR) in hospitals. The emerging trend in the medical practice is to move from ad hoc and subjective decision-making to evidence-based medicine (EBM) (J-Ath-Training, 2004 Jan-Mar). EBM is aimed at bringing Big Data to the healthcare consumer. Leading organizations such as IBM are playing a lead role in such initiatives. Analysis of health data reveals trends and knowledge, which at times, reveals contradictions to medical assumptions, which in turn, causes a shift in ultimate decisions that would better serve both patients and healthcare enterprises. In today's data-driven age, healthcare is making a shift from opinion-based decisions to informed decisions based on data and analytics. There are a few more drivers that cause growing complexity and abundance of healthcare data, which have, in turn, an advanced role of data analytics in healthcare: (1) new technologies such as capturing devices, sensors, and mobile applications, (2) the ease of use and the drop in cost of collection of genomic information, (3) the rise in digital forms used for patient social communications, and (4) the greater accumulation of medical knowledge/discoveries.

The Role of Data Analytics in Healthcare

As mentioned earlier, healthcare has three major challenges, cost, quality, and reach, i.e., (1) how to contain the cost of treatment and operative costs, (2) how to improve the quality (improve diagnostics, better outcomes from patient treatments), and (3) how to extend the reach of healthcare services.

Leveraging Big Data to the Benefit of Healthcare

A number of possibilities exist for reaping the benefits of "Big Data" for healthcare. A major one is about *increasing adoption rates for EMR i.e. electronic medical records*. EMR is one of the major challenges in healthcare domain; there are nine elements of a "usable" EMR: (1) simplicity, (2) naturalness, (3) consistency, (4) feedback, (5) effective use of language, (6) efficient interactions (among the relevant stakeholders), (7) effective information presentation, (8) preservation of

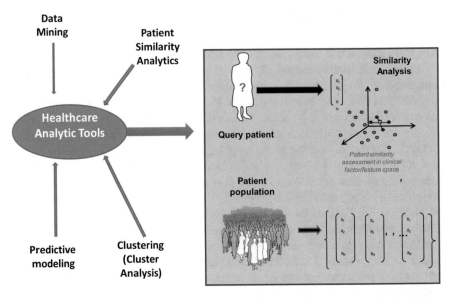

Fig. 2.2 Big Data analytics in healthcare (Source: Adopted from "Visual Analytics for Healthcare: Big Data, Big Decisions," David Gotz Healthcare Analytics Research Group IBM T.J. Watson Research Center)

context, and (9) minimum cognitive load. EMR (electronic medical records) are used for enhancing evidence-based medicine practices and for exploiting medical data to intervene earlier. Figure 2.2 presents a schematic view on the role of data analytics tool in healthcare for patient treatments.

Regarding "early intervention," a number of challenges exist. Here are some aspects of those challenges – (i) the scale is large (typically, up to tens of millions of patients), (ii) high dimensionality is involved (there are thousands of dimensions spanning many years), and (iii) the reference data required is semi-structured. For example, there are clinical notes which need to be integrated as EHRs i.e. electronic health records and EMRs i.e. electronic medical records, medical imaging, and medical codes. In addition, these data are geographically distributed: involving multiple providers and multiple representations.

A number of other challenges exist: there is a need to make critical decisions, or else it may literally mean life or death. There are no clear "right answers," and medical evidence is often ambiguous. There is limited time to manage complexity and multiple granularity. Domain experts are humans and, as such, they are limited by the best of their abilities. Patients show up in periodic visits and for each patient, different data gets generated. There are uncertainties as well: subjective errors, data entry errors, bias in billing, and incomplete information. Often some items could be missing from the medical record; this is the challenge of there not being integrated EMRs (electronic medical records).

Using Green IT and Data Science to Make Healthcare Green

Server virtualization is one of the most common ways for organizations to go green. Server virtualization in the data centers of the healthcare industry provides an opportunity for reducing the number of physical servers used. This, in turn, brings down physical hardware costs and the data center's carbon footprint. A report by the 451 Group titled "Eco-Efficient IT" (J-Ath-Training, 2004 Jan–Mar) found that each server eliminated through virtualization can reduce power consumption in a data center by up to 400 watts, which is the equivalent of about $380 per year, per server.

Desktop virtualization not only lowers energy costs; it can also increase productivity and decrease capital expenses on PC hardware. Information technology supports thin-client computing by centralizing management of all user desktop environments on a single platform. Forrester Research published a report in 2009, based on a research project they had undertaken to compare thin clients to desktops. It was found that thin clients consume between 5 and 60 watts per device, compared to the 150–350 watts used by a desktop PC.

Virtual collaboration provides a number of opportunities for "greening" of healthcare – IT supports web conferencing and instant messaging. There also are many software applications to promote virtual collaboration. By using virtual meetings and other collaborative efforts, travel and other expenses are cut dramatically. The effective use of virtual collaboration increases efficiency and enables employees to access information and applications anywhere at any time; this goes well with the adoption of EMR and EHR in hospitals. As an added benefit, using this technology can increase productivity and teamwork.

Here are some specific ways to use Green IT and Data Science to make healthcare green:

Implement Efficient Applications and Deduplicate Data

Software and application efficiency can be very significant for green IT. The author's recent experience shows that the procedure for creating a data warehouse report was reduced from 8 h to 8 min merely by changing the Oracle data warehouse search procedure (e.g., not searching the entire database each time when only a much smaller search is required). During the 8 h required to create the report, the large server was running at near peak capacity. Even if application inefficiency was created and fixed many times over the history of programming, what about the cases where a few application efficiencies would make an application run 20% faster"? That 20% more efficient application would also result in 20% lower energy use. The steps required to improve application efficiency by a few percent are often not easy to determine. However, the added incentive of saving energy, while making the application run faster, is a significant plus.

Data-storage efficiency, such as the use of tiered storage, is also very significant. Data deduplication (often called "intelligent compression" or "single-instance storage") is a method of reducing storage needs by eliminating redundant data. Only one unique instance of the data is actually retained on storage media, such as disk or tape. Redundant data are replaced with a pointer to the unique data copy. For example, a typical email system may contain 100 instances of the same 1-megabyte (MB) file attachment. If the email platform is backed up or archived, all 100 instances are saved, requiring 100 MB storage space. With data deduplication, only one instance of the attachment is actually stored; each subsequent instance is just referenced back to the single saved copy. In this example, a 100 MB storage demand can be reduced to only 1 MB.

Data deduplication offers other benefits too. Lower storage space requirements will save money on disk expenditures. The more efficient use of disk space also allows for longer disk-retention periods, which provides better recovery time objectives (RTO) for a longer time and reduces the need for tape backups. Data deduplication also reduces the data that must be sent across a WAN for remote backups, replication, and disaster recovery.

Data deduplication uses algorithms to dramatically compress the amount of storage space needed. Many organizations are dealing with increased scrutiny of electronically stored information because of various regulations; this need to preserve records is driving significant growth in demand for storing large sets of data. Depending on the type of information being compressed, deduplication can enable a compression rate of between 3:1 and 10:1, allowing healthcare facilities to reduce their need for additional storage equipment and associated tapes and disks. Many healthcare facilities are already using the technology.

Six Sigma to Help Make Healthcare Green

Six Sigma is a set of techniques and tools for process improvement. There techniques, originally developed by Motorola in the 1980s are being used by many industries throughout the world. The Six Sigma steps applicable for green IT and sustainability (Franchetti, 2013; Johnson & Johnson, 2012) are:

1   Define
2   Measure
3   Analyze
4   Improve
5   Control

These five steps are similar to the five-step process that IBM has used for several years for creating energy-efficient "green" data centers: (1) diagnose, (2) manage and measure, (3) use energy-efficient cooling, (4) virtualize, and (5) build new or upgrade facilities when feasible. Details on using the Six Sigma steps for green IT are described below.

1. Define the opportunities and problems (diagnose)

The step here is to do a data-center energy-efficiency assessment. The assessment should include a list of unused IT equipment that can be turned off. In addition, the diagnostic phase can help encourage organizations to retire unused software applications and focus on adopting more effective software that requires fewer CPU cycles. A typical x86 server consumes between 30% and 40% of its maximum power when idle. IT organizations should turn off servers that do not appear to be performing tasks. In case of complains, organizations should look into whether the little-used application can be virtualized. Look for electric utilities that offer free energy audits.

2. Measure ("you can't manage what you can't measure")

Many hardware products have built-in power management features that are never used. Most major vendors have been implementing such features for quite some time. These features include the ability of the CPU to optimize power by dynamically switching among multiple performance states. The CPU will drop its input voltage and frequency based on how many instructions are being run on the chip itself. These types of features can save organizations up to 20% on server power consumption.

3. Analyze IT infrastructure with energy-efficient cooling

Although many data centers may use hot aisle/cold aisle configurations to improve cooling efficiency, there are also some small adjustments they can make. Simple "blanking panels" can be installed in server racks that have empty slots. That's a great way to make sure the cold air in the cold aisle does not unintentionally mix with hot air in the hot aisle, wasting cooling and possibly failing to adequately cool the servers. Organizations should also seal cable cutouts to minimize airflow bypasses. Data organizations should consider air handlers and chillers that use efficient technologies such as variable frequency drives which adjust how fast the air-conditioning system's motors run when cooling needs dip.

4. Improve (virtualize IT devices – use cloud when feasible)

Virtualization continues to be one of the hottest topics in the discussion of green data-centers. Many current server CPU utilization rates typically hover between 5% and 15%. Direct-attached storage utilization sits between 20% and 40%, with network storage between 60% and 80%. Virtualization can increase hardware utilization by 5–20 times and allows organizations to reduce the number of power-consuming servers. Cloud computing is the "ultimate" in virtualization and we discuss virtualization and cloud computing in more detail later in this paper.

5. Control (continue to measure and manage and upgrade as necessary)

Going green is easiest if you are building a new data center. First, you make a calculation of your compute requirements for the foreseeable future. Next, you plan a data center for modularity in both its IT elements and its power and cooling. Then you use data-center modeling and thermal assessment tools and software – available from vendors such as APC, IBM, and HP – to design the data center. The next step is to procure green from the beginning – which partly means buy the latest equipment and technologies such as blade servers and virtualization.

Once you have the equipment, you integrate it into high-density modular compute racks, virtualize servers and storage, put in consolidated power supply, choose from a range of modern cooling solutions, and, finally, run, monitor, and manage the data-center dynamics using sensors that feed real-time compute, power, and cooling data into modern single-view management software that dynamically allocates resources.

Chapter Summary and Conclusions

The future of the healthcare industry is a major concern worldwide. As discussed in this chapter, the climate crisis and healthcare crisis are connected. Green healthcare can contribute to solving the climate crisis by adopting green IT practices. Success with healthcare initiatives to reduce cost and improve services will be significantly influenced based on the success of healthcare green IT initiatives. As also discussed in this chapter, emerging technologies such as mobile computing and cloud computing hold great promise to reduce costs and significantly improve services in the healthcare industry. The continued significant use of electric energy for the IT infrastructure used to support the healthcare industry has increased pressure on the industry to support green IT initiatives and overall sustainability. Healthcare green IT efficiency improvements must be made in compliance with the expanding regulations to protect patient privacy (Godbole, 2009).

Going forward, we have a role in helping improve the outlook for healthcare by contributing to IT infrastructure electric energy sustainability, data protection, and the continued improvement in cloud computing for IT cost reduction along with improved data protection.

Additional Reading

Field, R. I. *Why Is Health Care Regulation So Complex?* Accessed at the URL http://www.ncbi.nlm.nih.gov/pmc/articles/PMC2730786/.

Health Care's Climate Footprint. https://noharm-global.org/issues/global/health-care%E2%80%99s-climate-footprint.

"The Growing Importance of More Sustainable Products in the Global Health Care Industry", research study by Johnson & Johnson -http://www.jnj.com/wps/wcm/connect/ef4195004cca13b8b083bbe78bb7138c/JNJ-Sustainable-Products-White-Paper-092512.pdf?MOD=AJPERES.

mHealth: From Smartphones to Smart Systems, By Rick Krohn, MA, MAS, David Metcalf, PhD, published in 2012 by Healthcare Information and Management Systems Society (HIMSS), ISBN 978-1-938904-19-6.

Special Delivery: An Analysis of mHealth in Maternal and Newborn Health Programs and Their Outcomes Around the World, Maternal and Child Health Journal July 2012, Volume 16, Issue 5, pp 1092–1101, Stan Kachnowski, Tigest Tamrat, accessed on 23rd October 2016 at the URL as below http://link.springer.com/article/10.1007/s10995-011-0836-3.

An assessment of m-Health in developing countries using task technology fit model, FACULTY OF COMMERCE – PAPERS (ARCHIVE) accessed at http://ro.uow.edu.au/commpapers/3127/. Accessed on 23rd October 2016.

Mobile for Health (mHealth) in Developing Countries: Application of 4 Ps of Social Marketing, Dr. Dhanraj A. PATIL, *Journal of Health Informatics in Developing Countries* [JHIDC], Vol 5, No 2 (2011). Accessed on 23rd October 2016, at the URL quoted below http://jhidc.org/index.php/jhidc/article/viewArticle/73.

e-Health in India. Accessed on 23rd October 2016 at the URL mentioned below http://india.nlembassy.org/binaries/content/assets/postenweb/i/india/netherlands-embassy-in-new-delhi/import/e-health-in-india.pdf.

The mHealth Case in India Telco-led transformation of healthcare service delivery in India, Wipro White Paper. Accessed on 23rd Oct 2016 at the URL mentioned below http://www.wipro.com/documents/the-mHealth-case-in-India.pdf.

The Advent of Mhealth in India, Priyankasingh Veer Chandra Singh Garhwali Govt Medical Science And Research Institute, Pauri Garhwal, Srikot, Paper. Accessed on 23rd Oct 2016 at the URL – http://www.alliedacademies.org/articles/the-advent-of-mhealth-in-india.pdf.

Chapter 3
Healthcare and Green IT

Healthcare accounts for 8% of the U.S. carbon footprint

— University of California Hospitals Report

Green practices in the healthcare domain are the need of the hour considering the *climate change* challenge. A discussion about the importance of "green healthcare" for our planet is presented in Chap. 1. As mentioned there, being able to extend the *REACH* of healthcare services is one of the three challenges being faced by healthcare worldwide; the other two being "how to reduce the *COST* of healthcare services delivery" and "how to improve the *QUALITY* of healthcare services." The discussion in this chapter is important from the REACH as well as COST perspective. The first part of this chapter presents a discussion about "telemedicine" from the perspective of "greening" of the healthcare industry as well as reducing the costs of healthcare service delivery. The second part of the chapter is about the role of "virtualization" in that context. We begin by understanding the concept of green IT in the context of healthcare. Next, we introduce and explain some of the important terms related to "telemedicine," and in the next section, a discussion about the term "telemedicine" is presented.

The GREEN Context for Healthcare (IT and non-IT)

It was mentioned in Chap. 1 that hospitals form the largest segment of healthcare industry. There are many ways in which hospitals can go "green." Some of the ways of going "green" are related to the IT (information technology) aspect of hospitals, and there are some that are "non-IT" aspects.

Many hospitals, which have a commitment to "go green," consider reworking their *energy consumption* to modify their energy usage patterns; this indeed can be considered as a green IT aspect because the IT infrastructure of a hospital consumes electrical energy, although not to the degree of HVAC usage. The disposal of solid

© Springer International Publishing AG, part of Springer Nature 2018
N. S. Godbole, J. P. Lamb, *Making Healthcare Green*,
https://doi.org/10.1007/978-3-319-79069-5_3

(non-electronic) and electronic waste generated in hospitals (Information Resources Management Association (USA), 2011) and monitoring the use of chemical substances to reduce their impact on the environment is also one of the ways of making hospitals green. Through their sustainability initiatives, hospitals can save energy (Health Research & Educational Trust, 2014; PRACTICE Green Health, 2014). However, it is not easy to know where and how a hospital can start with its green initiatives (Boone, 2012). For many hospital practitioners, the breadth of sustainability issues can seem overwhelming – especially in hospitals that have the crunch of staff members, time, and budget – and as such, it becomes difficult for them to address these issues. Some practitioners whom we interviewed suggested that hospitals establish an environmentally preferable purchasing team to create environmental benchmarks and make sound purchasing decisions.

There are five recommended ways that hospitals can get maximum green results with little effort:

- *Consider local supply chains (if possible), i.e., go local*: Every day a huge quantity of food is served by the cafeterias operating on hospital campus. The source, i.e., the supply chain of that food, can have a dramatic effect on the hospital's environmental impact. Hospitals can contract with local suppliers and local supplying organizations to use more locally grown, fresh produce, which cuts down on gasoline involved in the transportation of those goods to the hospital as well the energy used to ship and refrigerate food coming from faraway places. In addition, locally sourced vegetables, for example, can be healthier for the patients. Hospitals can also work with local composting companies to recycle food waste that local farms will later use as fertilizer.

- *Look into water conservation*: One leading example comes from Virginia Mason Medical Center in Seattle (EnviroMason, 2016). It is a nonprofit hospital, and it saved more than 6 million gallons of water per year by making a number of changes in their treatment plans. They replaced the linear accelerator used in radiation therapy with a better model; they also replaced bathroom toilets, faucets, and showers with more efficient alternatives; and they purchased high-efficiency dishwashers. Under the scenario of large-scale water usage (like in hospitals), reducing water per flush or shower can make a huge difference in water consumption.

- *Try consuming less energy*: While it is not so easy for hospitals to reduce energy use and carbon output, it is not impossible. Consider Greenwich Hospital in Connecticut (Greenwich Hospital, 2014–2016) as another leading example – they saved more than 1.7 million kilowatt hours, cutting its electric costs by $303,000 per year and reducing its overall energy consumption by 35%. The hospital achieved this by reprogramming its heating and cooling plants. They also reengineered their air handling systems and updated its light bulbs. The fact that Greenwich Hospital made back its money within 6 months can be a motivator to other hospitals thinking "green."

- *Do rethinking on greener waste disposal methods*: According to Practice Greenhealth, hospitals in the United States (Greenwich Hospital, 2014–2016) produce more than 5.9 million tons of waste annually. The variety of waste they generate is a big challenge for hospitals – it makes environmentally friendly disposal difficult. For example, regulated medical waste must be disinfected before sending it to a landfill. This is not only a good practice but at times, a regulation to make sure that there is no environmental contamination. Disinfection methods, such as incineration, are not only energy intensive but also undesirable because these methods release noxious fumes. Processes such as chemical treatment, autoclaving, and microwaving can differ depending on how environmentally friendly they are. Hospitals ought to check with their waste disposal providers as to what type of energy and chemical use they have in their disinfection process and consider switching if they can find a greener company.

- *Consider chemical safety*: Under certain conditions, the chemicals used in hospitals are dangerous. Some of them are the electronic waste items in hospitals. The dangerous items include, for example, LCD displays, fluorescent lamps, CRT monitors, flame retardant mattresses, wheelchair cushions, and baby bottles. Some of those products can contain hazardous chemicals if purchased from the wrong manufacturer. Hospitals should think about the "source" from where they purchase products and must also think about proper recycling of toxic goods such as batteries.

Sustainability thinking for healthcare (mainly for hospitals) comes from the "green-triad" that embraces (1) green infrastructure, (2) green IT, and (3) green medical technology (Godbole & Lamb, 2013):

1. *Green infrastructure*

   - Cost savings through the use of optimized and modernized heating, ventilation, and cooling
   - The use of optimized illumination control systems for greater comfort and energy savings
   - Deployment of customized solutions, services, products, and technologies to optimize the life-cycle performance for hospitals with a view to achieve maximum energy savings, performance, and sustainability without losing safety and comfort
   - Less travel by using modern communications systems, e.g., videoconferencing, telemedicine, etc.
   - Green patient multimedia infotainment
   - Desktop phones with reduced energy consumption
   - Healthcare waste and eWaste, i.e., electronic waste (electronic and electrical equipment) disposal through 'reduce-reuse-recycle' principle

2. *Green IT*

   - Green transformation of healthcare data centers by using consolidation and virtualization technologies

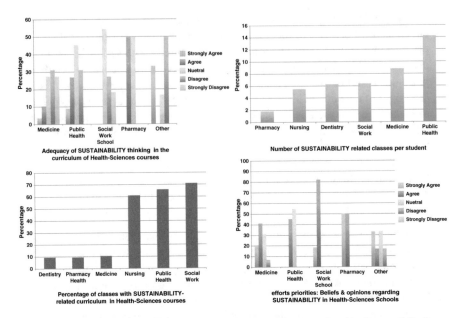

Fig. 3.1 "Green thinking" in health sciences. (Courtesy: Jake Vogt, Jennifer Power, Julia Bucy, Greening UW Health Sciences, UW Sustainability Office)

- Improving data-center energy efficiency file server storage reduction and usage of groundwater for air cooling in data centers
- Green IT architectures with desktop and server virtualization

3. *Green medical technologies*

- Energy-efficient medical technologies
- Making operation of medical technologies environmentally friendly

Given the needs of our planet, green thinking needs to be part of pervasive thinking in all domains of social networks as well as global organization. In this context, Fig. 3.1 shows the status on this scenario.

Mobile Devices and Telemedicine in Healthcare Domain

There are a number of definitions of telemedicine. According to the World Health Organization (WHO), telemedicine is defined as, "The delivery of healthcare services, where distance is a critical factor, by all healthcare professionals using information and communication technologies for the exchange of valid information for diagnosis, treatment and prevention of disease and injuries, research and evaluation, and for continuing education of healthcare providers, all in the interests of advancing the health of individuals and their communities" (Indian Telemedicine Network,

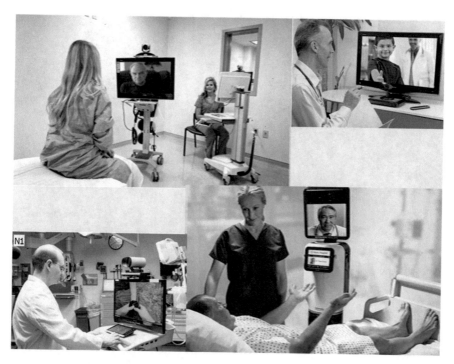

Fig. 3.2 Telemedicine at work: 1

n.d.). In broad terms, telemedicine can be defined as the *use of information technology to deliver medical services and information from one location to another.* Figures 3.2 and 3.3 present some illustrations involving telemedicine scenarios.

There are other terms often heard in synonymous use with the term "telemedicine," and they are "tele-health," "mHealth," and "eHealth." *mHealth* means "mobile health," and it points to the use of "mobile technologies" in the domain of healthcare service delivery (Preuveneersa, Berbers, & Joosen, 2013). Tele-health is the use of electronic information and telecommunication technologies to support long-distance clinical healthcare, patient and professional health-related education, public health preparedness, public health, and health education. *mHealth* (also written as m-health or mobile health) is a term used for *the practice of medicine and public health, supported by mobile devices* (Wikipedia, 2016). The term is commonly used in the context of using mobile communication devices, such as PDAs (personal digital assistants), mobile phones, and tablet computers or a combination thereof, for delivering health services and managing health information. *eHealth* means "electronic health," a term that points to "electronically delivered" healthcare services. eHealth is the term used for referring to health information technology (HIT). It has the potential to make healthcare a greener business. eHealth is a broad term that includes work within a healthcare organization where ICT (information and communication technology) is of central importance. There are a number of aspects to eHealth, including electronic medical records (EMRs), home monitoring of

Fig. 3.3 Telemedicine at work: 2

patients' vital parameters using mobile technology, as well as electronic health surveillance systems.

There are two broad ways for categorizing eHealth applications: (1) the use of
distance-spanning technology for healthcare through telemedicine/virtual visits
(i.e., remote diagnostics, tele-homecare, video consultations) and (2) the use of
electronic documentation of health services (e.g., electronic health records, i.e.,
EHRs, electronic prescriptions, surveillance systems). The term eHealth often is
used interchangeably with telemedicine, telecare, or tele-health. These terms are
used to basically describe the utilization of electronic communications to transfer,
share, and exchange medical information from one entity or site to another.
Examples include expert consultations using teleradiology, at-home speech therapy
using the Internet, and video technology. Interestingly, the market size of global
teleradiology was valued at USD 1636.6 million in 2015 and is expected to grow at
a CAGR of over 19.0% from 2016 to 2024 (Grand View Research, Inc, 2016).

All around the world, especially in the developed countries, there has been a
tremendous growth in the use of mHealth activities. According to a 2011 global
survey, undertaken by the World Health Organization (WHO), it was found that
mHealth initiatives have been established in many countries. The survey noted a
variation in the adoption levels of mHealth services. The survey revealed that the
creation of health call centers to respond to patient inquiries was the most common

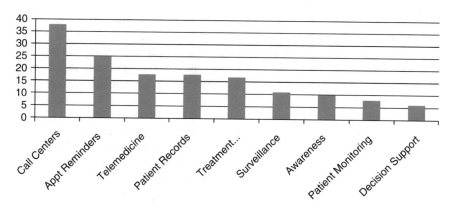

Fig. 3.4 Global mHealth initiatives around the globe; rate of adoption. (Courtesy Darrell M. West, Center for Technology Innovation at the Brookings Institution)

activity for mHealth (see Fig. 3.4). The next most common mHealth activities were found to be the use of SMS for appointment reminders. The other mMedicine activities found to be in use were using telemedicine services, accessing patient records, treatment compliance measurement, treatment surveillance, raising health awareness, patient monitoring, and decision support systems for physicians. Figure 3.4 presents this information about the adoption of mHealth worldwide (West 2012b, West 2010). The rate of adoption of mMedicine or telemedicine is the lowest in Africa countries and the highest in North American countries. In the South American countries and the countries in Southeast Asia, the rate of adoption is higher than that in African but lower than that in the North American countries (West, 2012a).

In recent years, the term mobile health (mHealth) is becoming increasingly popular with the advent of mobile technology with network interfaces developed to support mHealth applications – for example, the vast rise in the availability of mobile phones and tablets. There has been a phenomenal advance in the use of mobile technology. Cisco Visual Networking Index indicates that mobile data traffic will grow at a CAGR of 53% between 2015 and 2020, reaching 30.6 exabytes per month by 2020. It is estimated that "global mobile data traffic increased 18-fold between 2011 and 2016." By the end of 2020, it is predicted that the number of mobile devices in use around the world will be ten billion! This is not the only aspect of "Big Data." Along with 3G and 4G, these advances have had a huge impact on many walks of life, and one of those impacts will be or already is the way in which healthcare services are delivered. Recall that in Chap. 1, we mentioned the triple challenge for worldwide healthcare, i.e., (1) minimizing the *cost*, (2) improving the *quality* of healthcare, and (3) extending the *reach*, i.e., the *access*, to healthcare. Now mobile technology is all set to change the way how healthcare is delivered: the cost of healthcare, the quality of the patient experience, and improving the reach through the use of telemedicine (mHealth, eHealth).

In the domain of chronic disease management, mobile technology is tremendously promising (Laranjo et al., 2015; Blake, 2008; Scherr et al., 2009; Santoro,

Castelnuovo, Zoppis, Mauri, & Sicurello, 2015). In many areas of the world, the greatest healthcare challenge is the management of chronic diseases. Mobile technology-based remote monitoring devices enable patients with serious problems to record their own health measurements and send them electronically to physicians or specialists. Under such a scenario, the result is that it keeps the patients out of doctors' offices for routine care and thereby helps to reduce healthcare costs; recall that cost is one of the worldwide challenges in healthcare mentioned in Chap. 1. One analysis revealed that in the United States, remote monitoring technologies had the potential to save as much as $197 billion over the next 25 years (Brookings Institution, 2014; West, 2010). As a case in point, consider this: real-time management matters a lot in the case of chronic diseases. For example, for patients with diabetes, it is crucial that patients monitor their blood glucose levels and properly adjust their insulin intake to proper levels.

Meeting the doctor in person was easier in the "old days." Patients used to visit a doctor's lab or medical office and get the prescribed test, and then they would wait for results to be delivered. That process was not only expensive but time-consuming. It also was inconvenient for all involved. One of the factors that drive up medical costs is getting regular tests done for things like diabetes and other conditions. And this is not all; electronic health record (EHR) and/or electronic medical record (EMR) adoption rate is not yet very high (Exhibit 3.1 explains the difference between an EHR and EMR). Therefore, one of the barriers to cost containment and quality service delivery has been the continued reliance in many locales on paper-based medical systems, i.e., not using the EHRs or the EMRs. Manual forms are used by physicians to prescribe medicine, lab tests are reported on paper, medical records are stored in filing cabinets, and insurance claims get paid through reimbursement requests sent through the mail. In one of the large hospitals surveyed in India (Pune city), we found this is to be the situation. One cannot imagine a costlier way to run a healthcare system in this manner, given that today we are in a digital world. The potential for healthcare cost saving is supported by other similar studies (Garber, 2000; Garber & Phelpsc, 1997).

The introduction of PDAs (personal digital assistants), smartphones, and tablet computers, as well as other mobile computing devices has greatly impacted medicine. Healthcare professionals now use tablet computers or smartphones (instead of a pagers), cell phones, and PDAs to accomplish most of their professional tasks. Smartphones and tablets offer a range of benefits to healthcare professionals (HCPs); they combine both computing and communication features in a single device that can be held in a hand or stored in a pocket, thus allowing easy access and use at the point of care. The new mobile device models, in addition to voice and text, offer more advanced features, such as web searching, global positioning systems (GPS), high-quality cameras, and sound recorders. These advanced technological features, as well as powerful processers and operating systems, large memories, and high-resolution screens, have converted mobile devices into handheld computers.

Mobile technology in chronic disease management empowers elderly persons, especially those living by themselves and expectant mothers. It helps by reminding people to take medication at the prescribed time intervals. It addresses the 'reach'

Exhibit 3.1: EMR and HER: The Two Related but Different Things

Often, the two terms EMR (electronic medical record) and EHR (electronic health record) are used as interchangeable terms. However, we need to clearly distinguish between these two terms because an EMR is not the same as EHR. They describe concepts that are completely different, and both (i.e., electronic medical records and electronic health records) play a crucial role in the success of healthcare industry's goals (1) to improve the quality, (2) to improve patient safety, (3) to improve efficiency of patient care, and (4) to reduce healthcare delivery costs.

Basically, an electronic medical record (EMR) is simply a digital version of a paper chart that contains all of a patient's medical history from one clinical practice. An EMR is mostly used by providers for the purpose of diagnosis and treatment. An EHR, on the other hand, is a subset of EMR. EHRs (electronic health records) rely on EMRs (electronic medical records) being established, and at the same time, EMRs will never reach their full potential without the availability of interoperable EHRs. Therefore, it is important to understand the differences between them.

The EMR is the *legal record* created in hospitals and ambulatory environments and serves as the source of data for the EHR. The EHR represents the infrastructure for easily sharing medical information among stakeholders (patients/consumers, healthcare providers, employers, and/or payers/insurers, including the government) and to have the continuous availability patient's updated information throughout the various modalities of care delivered to that individual.

However, healthcare service provider organizations must implement complete EMR solutions before they can move to effective EHR environments. In spite of great efforts being done in that directions, only a few hospitals have EMR solutions that are effective enough to reduce medical errors or improve the quality and efficiency of patient care (Brooks & Grotz, 2010).

According to HIMSS (Healthcare Information and Management Systems Society) Analytics, electronic medical record is an application environment composed of the clinical data repository, clinical decision support, controlled medical vocabulary, order entry, computerized provider order entry, pharmacy, and clinical documentation applications. In such an environment, the patient's electronic medical record (EMR) gets supported across inpatient and outpatient environments, and healthcare practitioners use it to document, monitor, and manage healthcare delivery within a care delivery organization (CDO). The data in the EMR is the legal record of what happened to the patient during their encounter at the CDO and is owned by the CDO. Table 3.1 further explains the difference between the two important terms – the EMR and the EHR.

To understand what an EHR (electronic health record) contains, the URL mentioned below (accessed on November 2, 2016) can be visited.

(continued)

https://www.healthit.gov/providers-professionals/faqs/what-information-does-electronic-health-record-ehr-contain

The contents of digital medical records can be seen at the URL mentioned below (accessed on November 2, 2016). http://www.medicalrecords.com/what-is-an-electronic-medical-record

Table 3.1 The difference between an *EMR* and an *EHR*

Electronic medical record (EMR)	Electronic health records (EHRs)
An EMR is the legal record of the care delivery organization (the CDO)	Subset (i.e., CCR, Continuity of Care Record, or CCD, Continuity of Care Document) of information from various CDOs where patient has had encounters
EMR is a record of clinical services for patient encounters in a care delivery organization	The EHR is owned by patient or stakeholder
The CDO owns the EMR	EHRs are with community, state, or regional emergence today (RHIO, Regional Health Information Organization) – or nationwide in the future
EMR systems are sold by enterprise vendors and installed by hospitals, health systems, clinics, etc.	EHR provides interactive patient access as well as the ability for the patient to append information
In case of EMR, a patient may have access to some results info through a non-interactive portal	EHR is connected by NHIN (National Health Information Network)
An EMR does not contain information about interactions with other CDO (it is a record that is contained within a given CDO only)	

problem by extending medical services to underserved and/or inaccessible areas. The results of all this are the improvement of health outcomes and the improvement of medical system efficiency. One illustration comes from Mexico, a country where diabetes is the biggest chronic healthcare problem. In Tijuana, Mexico, healthcare workers work with a mobile application for interacting with patients. Participants are trained to use a mHealth application to access videos. The participants also receive health education on diabetes self-management and how to send interactive questionnaires to healthcare workers. 3G mobile phones are used to send patients' responses, making them instantly available in a database for the healthcare providers who review and follow up. Alerts are sent to patients when they report significantly out of range indicator levels or when the system identifies lack of adherence to appointments, tests, or classes.

In the previous section, we addressed the empowering role of mobile technology in the healthcare domain. We explained the role of telemedicine in healthcare, i.e., the "why" and "the benefits." Let us now understand telemedicine or mobile medi-

Exhibit 3.2: "mHealth" and Mobile Devices in Healthcare: The World Is Changing

The term "mHealth" is used for the mobile devices supported practice of medicine and public health. The adoption of mobile technologies in every dimension of life has been phenomenal. In the span of two decades, ever-more sophisticated mobile technology has fundamentally altered the ways in which people communicate and conduct business. The disruptive power of these new technologies and the accompanying waves of innovation they have sparked are transforming the healthcare industry, propelling stakeholders to reassess and repurpose how they provide services. In the United States, the sales of mobile device have grown from 172 million in 2009 to 215 million in 2016 which represents a 25% increase, and the revenue earned from mobile data usage has risen from $35 billion in 2008 to $180 billion in 2016; it is a meteoric rise of 514% increase! By the year end 2016, worldwide, there were 10 billion mobile devices in use. Mobile device users who downloaded at least one mHealth application onto their smartphone doubled between 2011 and 2012. As for the future of the healthcare industry in the United States, mHealth is a key element of the industry's response to the market's quest for value. Figure 3.5 presents the history of smart mobile devices.

Interestingly, in the healthcare industry too, it is the consumers of the healthcare services who are driving the demand for mHealth technologies and applications. Deloitte Center for Health Solutions, during their 2012 Survey of US Healthcare Consumers, identified six segments that are unique in the healthcare industry: (1) casual and cautious, (2) content and compliant, (3) online and onboard, (4) sick and savvy, (5) out and about, and (6) shop and save. Of these six segments, interestingly, the third segment (online and onboard) is constituted by consumers who are more active compared with those in other segments, and the consumers in this segment show a strong preference to use online tools and mobile applications to assess providers and compare treatment options and provider competence. The consumers in this segment navigate the healthcare systems in a very different way:

1. They have the highest levels of trust in web sites to provide reliable information.
2. They make high use of online resources to look for information about health problems and treatment options.
3. They are interested in using price and quality information to compare health plans, providers, and treatments.
4. They make the highest use of electronic personal health records and concern about privacy and security.
5. They are most interested in videoconferencing with doctors (see Fig. 3.6), using self-monitoring devices that could send information electronically to doctor, and using health improvement tracking apps.

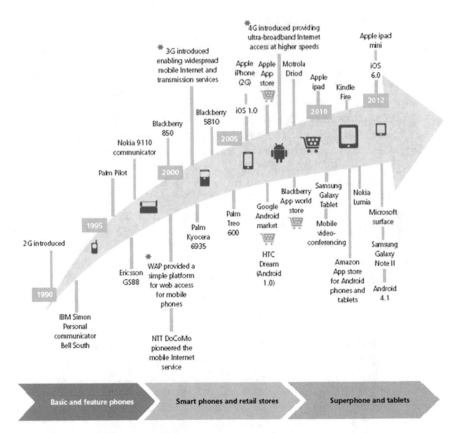

Fig. 3.5 History of smart mobile devices (Source: Brief history of Smart Phones. PC World Jun 18, 2010 http://www.guardian.co.uk/technology/2012/jan/24/smartphones-timeline)

cine in that context. There is a growing awareness about the need to reap the benefits of information technology (IT) and telemedicine to improve the health of our society and also to improve the "reach" of healthcare services delivery. Telemedicine makes it possible to provide access to healthcare for those living in remote rural areas or other areas with poor access to healthcare facilities. Mobile solutions in healthcare and the beneficial deployment of IT are of great relevance to developing countries (Cohen, 2013).

Similar to other industries, consumers in the healthcare industry are driving much of the demand for mHealth technologies and applications. The "cost" and "reach" dimension is mentioned in the beginning of the chapter; in that context, note that mobile apps have succeeded in increasing the flow of information, in helping to bring the costs down through better decision-making, reducing in-person visits, causing greater adherence to treatment plans and in improving satisfaction with the service experience. Thus, mHealth is becoming an extension of the intersection between technology and healthcare. As seen in Fig. 3.5, consumers tend to make a widespread use of mobile devices which makes it easier and faster to access health-

Fig. 3.6 An example of videoconferencing in telemedicine (Source: http://www.savvyhealth-carepr.com/wp-content/uploads/2017/01/telemedicine-1080x675.jpg)

care and creates opportunities to revolutionize the industry through high-quality and highly personalized care. In the next section, we discuss the benefits of telemedicine and mobile medicine. To help readers access more discussion about mobile health practices a number of resources are listed in the reference section at the end of the chapter.

Healthcare and the Benefits of Telemedicine and Mobile Medicine

The healthcare industry has been heavily impacted by mobile devices, although traditionally it has been slower as far as the adoption of new technologies is concerned. New possibilities emerge when a smartphone or tablet computer is put in the hands of a doctor, nurse, or administrator. However, it needs a good quality of back-end support. In the healthcare setting, the use of mobile devices holds the potential to enhance productivity, a promise to lower failure-to-respond rates, opportunities to increase information access, and communication. Some specific reasons that have made mobile technology popular over the last few years are (1) patient compliance to medication/treatment, (2) improved information access, and (3) being on top of the latest health trends in real time.

For populations living in isolated communities or remote and inaccessible regions, telemedicine and mobile medicine are most beneficial. They are most useful as a communication tool between a general practitioner (a GP) and a specialist available at another geographic location. The prefix "tele-" is used for medical specialties that make use of telemedicine. For example, teleradiology is telemedicine technology used by radiologists, telecardiology is telemedicine used by cardiologists, etc. These days, many healthcare professionals (HCPs) have started using mobile devices, and it has transformed many aspects of clinical practice. In healthcare settings, the use of mobile devices has become commonplace; it has resulted in a rapid growth in the development of medical software applications (apps) for mobile platforms. A number of applications are now available to assist HCPs with their important tasks, for example, health record maintenance and access to information and time management, communications and consulting, accessing medical references and information gathering, patient management and monitoring, clinical decision-making, and medical education and training. There are a number of benefits that mobile devices and apps provide to healthcare professionals. One of the most significant of them perhaps is the increased access to point-of-care tools. These are known to support better clinical decision-making and improved patient outcomes (Mickan, Tilson, Atherton, Roberts, & Heneghan, 2013 Oct 28).

The Barriers in the Use of Telemedicine and Mobile Healthcare

Some healthcare professionals (HCPs) remain reluctant to adopt the use of telemedicine and mobile healthcare practices despite the benefits that mobile devices offer, when used in healthcare (see Exhibit 3.2). There is a need to establish better standards and validation practices regarding mobile medical apps to ensure the proper use and integration of these increasingly sophisticated tools into medical practice. It is these measures that will raise the barrier for entry into the medical app market, increasing the quality and safety of the apps currently available for use by healthcare professionals. There is another aspect of the barriers in the use of telemedicine; for example, there are some factors impeding the growth of telemedicine including confidentiality and malpractice issues, technical advances, reimbursement, licensing, credentialing costs, cost-effectiveness, and legal issues (Medeiros de Bustos & Moulin, 2009).

There are a number of known disadvantages or limitations to telemedicine in rural areas. The first of these is the "cost" aspect; the overall cost of telecommunication system, especially data management system and practical training of medical professionals, is quite high. The second factor is about the human interactional aspect; as healthcare professionals' personal interaction with patients and other healthcare professionals decrease in a virtual environment, the risks of error in clinical services and treatments increase, especially if the services are delivered by less experienced healthcare professionals such as aides and assistants. Moreover, confidential and sensitive medical information can be leaked through insecure electronic systems. Telemedicine is highly dependent on the connectivity, which in turn, is impacted by the quality of IT infrastructure. As such, telemedicine might take a longer time due to the difficulties faced in connecting virtual communication due to low Internet speed or server problems. Moreover, the telemedicine system cannot provide immediate treatment, such as physical administration of medicines. Yet another issue in telemedicine could be concerning the possible low quality of health informatics records, like X-ray or other images, clinical progress reports, etc., and these may, in turn, cause the risk of faulty clinical treatment. Over and above these factors, note that telemedicine system requires tough legal regulation to prevent unauthorized and illegal service providers in this sector. mHealth practices have not been very successful in countries with a predominantly low and middle income population (Mechael et al., 2010).

The IT Infrastructure of Telemedicine

Telemedicine basically involves the use of telecommunication technology to provide medical information and services. It requires LAN (local area network) and the WAN (wide area network) with adequate bandwidth to support the transmission of images or video over the network. Bandwidth is the maximum amount of data (measured in bits per second) that can travel a communications path in a given time. Greater bandwidth makes it possible to transmit more data resulting in higher image and audio quality. If the bandwidth is not adequate, there will be delays in transmission of high definition graphic images, and this may impact the telemedicine consultation effectiveness.

The scenarios of telemedicine can be as simple as two healthcare professionals (HCPs) discussing a case over the telephone or as sophisticated as using satellite technology to broadcast a consultation among multiple HCPs at healthcare facilities in two or more countries using videoconferencing equipment. Thus, generally speaking, the practice of telemedicine refers to the use of ICT (information and communication technology) for the delivery of clinical care. Telemedicine is practiced using two concepts: (1) real-time (synchronous; it is live real-time and often video-based) and (2) store-and-forward (asynchronous). In an asynchronous mode, still images, video clips, vitals, radiology, etc. are captured and then transmitted to the physician. The concerned physician then reviews them his location (away from where they have been sent) and replies with diagnosis and treatment plan.

Real-time telemedicine practice could range from very simple to very complex. An example of a simple telemedicine task is a telephone call between two HCPs. An example of complex task would be using a remotely controlled robotic arm for performing a surgery with or without the guidance of an on-location surgeon. In real-time telemedicine, both the parties are required to be present at the same time, and it also requires a reliable communication link between them. Videoconferencing equipment is one of most common forms of ICT used in synchronous telecommunication. To support an interactive examination of a patient, peripheral devices can be attached to computers, to computer-aided diagnostics machines, or to the video-conferencing equipment. For example, a tele-otoscope allows a physician at a remote location to "examine" the patient by seeing the insides of their ears. With a tele-stethoscope, the consulting physician at another medical site is able to hear the patient's heartbeats.

Personal privacy issues may crop up in telemedicine practice (Ren, Werner, Pazzi, & Boukerche, 2010; Gritzalis, Lambrinoudakis, Lekkas, & Deftereos, 2005); and it involves a separate discussion about patients' "personal medical privacy" which is not within the scope of this book. In the reference section at the end, there are pointers to some useful resources about this topic.

"Store-and-forward medicine" involves the acquisition of medical data, for example, medical images, bio-signals, etc., and then its off-line transmission later to a doctor or a medical specialist. A properly structured EMR (electronic medical record) is used in off-line transfer of medical data. The store-and-forward method does not require that both the parties (the data transmitter and the data receiver) to be present at the same time, i.e., it supports "asynchronous" mode of data transmission. We find the use of "store-and-forward medicine" in medical areas such as dermatology, radiology, pathology, etc.

One of the most common uses of telemedicine today is sending X-ray images, CT scans, MRIs, etc. A large number of medical centers use "store-and-forward" technology for this purpose. Once the required IT infrastructure is installed by telemedicine professionals (e.g., radiologists), they can work at their own convenience, saving them physical trips to a hospital or a clinic.

Telemedicine brings a benefit to populations living in remote areas or in areas where the access to healthcare service is difficult. Thus, telemedicine is a useful tool when a general practitioner in a remote area needs to communicate to consult with a medical specialist in a faraway metropolitan area.

Another widely used technology in telemedicine is the two-way IATV (interactive television) – it is used when a face-to-face communication is required (see Figs. 3.2 and 3.3). Under this scenario of a telemedicine practice, the patient and his/her healthcare provider or a nurse or a telemedicine coordinator or a combination of these entities are at the originating site and the medical specialist at an urban medical center. Both ends, however, require videoconferencing facilities to make possible a real-time consultation. Unfortunately, in developing countries, the ICT infrastructure at the remote area often is either non-existent or not of good quality, marring the huge potential for telemedicine.

The Greening of Healthcare: Telemedicine and Green IT

As mentioned in the previous section, telemedicine involves the use of interactive healthcare over distance using information or telecommunications technology (ICT). Telemedicine leverages the knowledge of an expert to the point of care and allows that expertise to be customized for that patient. There are some studies conducted to measure the telemedicine on greenhouse gas (GHG) emissions (Masino, Rubinstein, Lem, Purdy, & Rossos, 2010). As per this study, approximately 360 kg of other air pollutant emissions was avoided. The GHG emissions due to energy consumption involved in videoconference units were estimated to be 42 kg of carbon dioxide equivalents emitted for the sample considered in the research study. The study shows that the overall GHG emissions associated with videoconferencing energy use are minor when compared with the energy avoided from vehicle use. The adoption of telemedicine services usually results in environmental benefits, and this works as an additional incentive over and above the improved patient-centered care and cost savings.

In another study (Yellowlees, Chorba, Parish, Wynn-Jones, & Nafiz, 2010), it has been argued that telemedicine and health information technology help save time, energy, raw materials (such as paper and plastic), and fuel, thereby lowering the carbon footprint of the health industry. It is further argued that by implementing green practices, for instance, by engaging in carbon credit programs, the health industry could benefit financially as well as reduce its negative impact on the health of our planet. Business operations that reduce their carbon emissions by implementing energy-saving practices can sell their carbon credits to companies that emit more carbon than permissible by their legally binding commitment. These carbon profits can be used for research in healthcare domain or can be used for providing healthcare to those who have not been served adequately. On the other hand, the savings could be used for green purchasing and to implement other carbon-reducing activities. There are several possible options for the American health industry to become greener and lower its carbon footprint while at the same time becoming more time- and cost-efficient.

In a study conducted in year 2010 (Masino et al., 2010), the reduction in greenhouse gas (GHG) emissions resulting from 840 telemedicine consultations completed in a 6-month time period was estimated. The estimation model considered was based on the impact of both vehicle and videoconferencing unit energy use on GHG emissions. In this study *cost* avoidance factors were also discussed. Travel distances in kilometers were calculated in this study, for each appointment using postal code data and Google Maps™ web-based map calculator tools. The study showed that an estimated 757,234 km were avoided including return travel, resulting in a GHG emission savings of 185,159 kg (185 mt) of carbon dioxide equivalents in vehicle emissions. Further it was found that approximately 360 kg of other air pollutant emissions were also avoided. Based on the sample considered for the study, it was estimated that GHG emissions produced by energy consumption for videoconference units were 42 kg of carbon dioxide equivalents emitted.

Interestingly, the study concluded that the overall GHG emissions associated with videoconferencing unit energy is minor when compared with those avoided from vehicle use. It was also concluded that in addition to improved patient-centered care and cost savings, environmental benefits provide additional incentives for the adoption of telemedicine services.

Therefore, from the "carbon-footprint" perspective of the healthcare sector, the finding of this study supports the fact that the healthcare sector is a significant contributor to global carbon emissions, partly due to extensive traveling by patients and health workers. This point is supported by yet another study (Holmner, Ebi, Lazuardi, & Nilsson, 2014), in which it was found that replacing physical visits with telemedicine appointments resulted in a significant 40–70 times decrease in carbon emissions. Factors such as network bandwidth, duration of meetings, and usage rates influence emissions to various extents. Telemedicine becomes a greener choice at a distance of a few kilometers when the alternative is transport by car.

Thus, we can say that telemedicine can make healthcare greener. The healthcare industry worldwide is said to be lacking environmentally sustainable practices. The environmental impact of healthcare practices has been largely disregarded due to fears of additional costs and regulations, ignorance, and ambivalence. The current healthcare practices continue to pollute the environment by requiring large amounts of travel and paperwork by both the patient and the clinician. Health information technology and telemedicine help save energy, time, raw materials (such as paper and plastic), and fuel, thereby lowering the carbon footprint of the health industry. By implementing green practices, for instance, the health industry could benefit financially as well as reduce its negative impact on the health of our planet; one way is by way of engaging in carbon credit programs. Companies that are able to reduce their carbon emissions by implementing energy-saving practices can sell their carbon credits to companies that emit more carbon than permissible by their legally binding commitment. These carbon profits can then be used for the purpose of research in the healthcare domain or to provide healthcare to those who are underserved. As an alternative, the savings could be used for green purchasing practices and to implement other carbon-reducing activities. There are numerous possible options for the health industry to become greener and to lower its carbon footprint while at the same time becoming more time- and cost-efficient.

Telemedicine Practices and mHealth Services in Developing Countries: The Indian Scenario

With regard to the status of telemedicine practice, there are big differences between developed and developing nations. Africa is said to have the lowest rate of mHealth adoption, while North America, South America, and Southeast Asia show the highest adoption levels. A number of countries have initiatives in the pilot stage or have informal activities that are underway and India is one of them. The biggest mHealth obstacles were competing priorities, budgetary restrictions, and staff shortages. Concerns over privacy and data security also were cited as barriers to effective implementation. Most countries are implementing mHealth through various types of PPP (public-private partnerships).

As far as India is concerned, it is not surprising that in the early years, one of the strongest indicators of both the awareness and use of mHealth services is monthly spending on mobile phone service. This indicator, however, does not hold any longer – when examining the major barriers to mHealth adoption, the differences between lower and high mobile spending individuals became statistically insignificant. Accordingly a survey report (Lunde, 2013), there are a number of interesting revelations: those with lower monthly mobile spending are significantly more likely to believe that mHealth services were easier to use than the services they are currently receiving. Given that one of the biggest problems in the low-cost and free public healthcare system is employee absenteeism, this finding should not be surprising. Further, it was found that consumers with low spending on mobile phone are price conscious. They are either unable or unwilling to pay for high-cost, facility-based private healthcare facilities. mHealth, thus, has potential to fill the gaps in the system by providing expanded access at a lower cost. This phenomenon is illustrated in Fig. 3.7.

In developing economies like India, there are several key drivers that promote mHealth as well as barriers that impeded the wide practice of mHealth. Although according to the study report quoted (refer to Figs. 3.7, 3.8 and 3.9), there was an overwhelming agreement that mHealth services are valuable, there are, at the same time, some deterrents to the widespread adoption of mHealth services in India; some of the important ones are (1) concerns about poor network coverage; to illustrate the point, consider this – a patient being monitored for a heart attack cannot be left at the mercy of an underdeveloped mobile network. (2) Another overwhelming concern expressed by Indian respondents to the survey was the security and privacy of healthcare information (refer to Fig. 3.8). Adding a relatively new technology, such as mobile phones and allied applications raises doubts in people's mind. (3) The perception that mHealth applications would be too complex is the third most cited barrier to the adoption of mHealth services in India. The summary of these barriers to mHealth is presented in Fig. 3.8; it shows percentage of respondents who "agree."

Overall, however, the survey revealed that one of the key results is that mHealth services are looked upon as a convenient way to access high-quality information (see Fig. 3.9). A significant lesson from this study is that the two biggest barriers to mHealth adoption are (1) reluctance to pay and (2) a lack of value perceived from mHealth. As of now, in India, high-quality healthcare services are not readily available. In such a scenario and keeping in mind the population living in Indian villages, mHealth can be considered as the service with potential to be both a substitute and a complement to the traditional, in-person healthcare services.

To conclude the discussion in this section about telemedicine and mHealth services in developing countries like India, we note a few elements related to mHealth in the Indian context. National health issues are relevant to rural Indians as much as

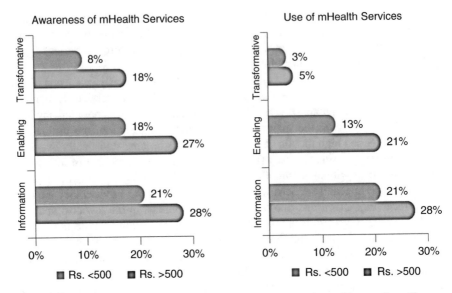

Awareness of mHealth Services   Use of mHealth Services

Fig. 3.7 Awareness about and use of mHealth service: type and monthly spending. (Source: Wipro and IAMAI Study, 2012)

Fig. 3.8 Barriers mHealth adoption in India. (Source: Source: Wipro and IAMAI Study, 2012)

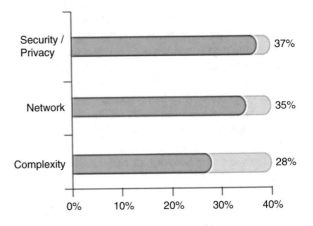

they are to urban Indians. There is room for care improvement throughout India. In India, infant mortality rate (double digit) is far higher than that in highly developed countries such as the United States. India has basic public health infrastructure issues. Hospital and physician supply is also limited in rural areas and remote areas. Leading cause of death and disability in India are chronic diseases. India's healthcare problems are complex and difficult to resolve. As such, there are vast opportunities for telecom, healthcare, and other organizations to design strategies for the delivery of mHealth services. mHealth services can play a key role in addressing India's chronic disease problem. In countries like India, healthcare access could be

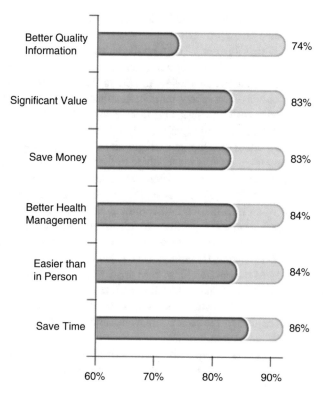

Fig. 3.9 mHealth drivers in India. (Source: Wipro and IAMAI Study, 2012)

expanded to rural communities to deliver healthcare servcies where there is limited capacity. Under such a scenario, a two-way model is likely to work for India, given the peculiarity of its health services scenario – (1) healthcare access would be expanded to rural communities and bring services where there is limited capacity and (2) care would be focused on the urban wealthy who have the ability to pay for costly monitoring services that stand to improve their health.

Telemedicine in Healthcare: The Challenges

As explained in the previous sections, a generic view of mHealth is that it is the delivery of healthcare services or information with a smartphone. mHealth services available globally have a great variation in their level of sophistication. mHealth services span the range from providing static information about a disease or illness to providing movement up the value chain by providing comprehensive healthcare management beyond the delivery that is done through the traditional face-to-face interaction with a healthcare provider.

Another aspect is that the lack of data on the benefits of telemedicine technologies has been a challenge to their adoption. There are challenges due to the heavily regulated nature of healthcare. Healthcare providers (HCPs) need to be licensed to provide care. Compliance is a heavy burden; healthcare institutions must comply with numerous regulations to operate. One more challenge is that the required policies and guidelines for licensing physicians to use telemedicine technologies for care delivery are still under development. Regulations are very stringent; physicians providing care using telemedicine must adhere to the same licensing regulations as physicians providing care in-person. There are also the technological factors to worry about when it comes to telemedicine practice. Seamless communication of IT systems is crucial when information and communication technology is used in the delivery of healthcare services. In addition to this aspect, healthcare service is characterized by the presence of multiple parties (care providers, medical insurance organizations, medical diagnostic laboratories, research organizations, other allied parties concerned with patient handling, etc.) apart from just the doctors. Sensitive medical information changes hands because each of these organizations has their own information systems that store patient health information. Adoption of telemedicine is hindered due to the lack of interoperability standards. There are also some cross-institutional factors.

A telemedicine network is typically characterized by the presence of multiple organizations that are formally and informally affiliated with one another. They come together engaged in the common objective of providing necessary care to remote patients. Payments or reimbursements (to the care providers for their services) are made by private insurance companies or through government programs. However, not all types of services are reimbursed. There are also compatibility issues regarding various forms/format and procedures associated with the payments and reimbursements. A key reason for the slow adoption of telemedicine is inconsistent standards for reimbursement mechanisms. Often there are no clear guidelines on reimbursement; as a result, healthcare providers and administrators often are reluctant to adopt telemedicine. The cost issue in healthcare is mentioned at the beginning of the chapter; it is applicable to telemedicine as well.

Telemedicine infrastructure involves costs of installation, operations, and maintenance and as a result is expensive. Due to a lack of wider adoption of telemedicine, the ROI (return on investment) becomes low, which in turn makes investment in such infrastructure less attractive. There are also other factors that impede the wide adoption of telemedicine: (1) high cost of technology and equipment, (2) lack of program funds, and (3) cost of ongoing technical support.

In decisions relating to technology adoption, stakeholders expect a strong evidence of benefits; this often is difficult to come by in telemedicine. It is, therefore, not surprising that there is low inclination toward investing in telemedicine technologies. At the same time, in order to be convinced about the benefits of telemedicine, payers and providers require this evidence base in terms of both reduction in cost and increase in care quality. Both cost and quality factors are mentioned in the beginning of this chapter.

Green Healthcare and the Role of Virtualization in Green IT

The most significant step most companies can make in their quest for green IT is in IT virtualization as was briefly mentioned in previous chapters. This section describes the significant concepts of virtual servers and virtual data storage for energy-efficient company data centers which carry over to healthcare systems. Much of the information in this section is from previous work on green IT for data centers (Lamb, 2009). The descriptions include VMware and other server virtualization considerations. In addition, the virtual IT world of the future that is fast becoming reality, via grid computing and cloud computing, is discussed. Although the use of grid computing and cloud computing in your company's data center for mainstream computing may be some ways off in the future, some steps toward that technology for mainstream computing within your company are here now. Server clusters via VMware's VMotion and IBM's PowerVM partition mobility are here now and being used in many company data centers. Both of those technologies are described in this section.

There are many aspects to IT virtualization. The chapter structure covers the rationale, server virtualization, storage virtualization, client virtualization, grid/cloud concepts, cluster architecture for virtual systems, and conclusions.

Over the past 30 or more years, data centers have gone from housing exclusively large mainframe computers to housing hundreds of smaller servers running versions of the Windows operating system or Unix/Linux operating systems. Often the smaller servers were originally distributed throughout the company, with small Windows servers available for each department in a company. During the past few years, for reasons of support, security, and more efficient operations, most of these distributed servers have moved back to the central data center. The advent of ubiquitous high-speed networks has eliminated the need for a server in the same building. These days, network access even to our homes through high-speed networks such as DSL and cable allows network performance from our homes or distributed offices to the central data center to be about equivalent to performance when your office is in the same building as the data center. The Internet was and remains the most significant driving force behind the availability of high-speed networks everywhere in the world – including to peoples' homes in most of the developed world. When we access a web site from our home, from the airport with a wireless connection, or from the countryside using a PDA or an air card with our laptop, we have a high-speed connection to a server in some data center. If the web site is a very popular site such as Google, the connection could be routed to any one of many large data centers.

When the distributed servers that had been in office buildings were moved in the past 10 years to centralized data centers, operations and maintenance became greatly simplified. With a company server at a centralized data center, you could now call the help desk on Sunday morning and find out why you had no access, and central operations could have a technician "reboot" the server if it had gone down. So the centralized data center provides many advantages – especially with high-speed networks that eliminate network performance concerns. However, with the rapid growth in servers used in business, entertainment, and communications, the typical data center grew from dozens of separate physical servers to hundreds of servers and sometimes to thousands. Purchasing, operating, and maintaining hundreds of separate physical servers became very expensive. The innovative solution was to consolidate perhaps ten of the separate servers into one bigger physical server but make it "appear" as if there were still ten separate servers. Each of the ten "virtual servers" could retain its own server name and its own Internet address (IP address) and appear – even to web developers – to be a separate physical machine (as it had been before becoming a virtual server). Costs go way down since one large physical box is much less expensive to buy than ten smaller physical boxes. Also, it's significantly less expensive to maintain and operate ("take care of") one big server than ten smaller servers. The analogy may be exaggerated – but it's a bit like taking care of one big house rather than ten separate smaller houses of the same total floor area and room count as the big house.

In simple terms, server virtualization offers a way to help consolidate a large number of individual small machines on one larger server, easing manageability and more efficiently using system resources by allowing them to be prioritized and allocated to the workloads needing them most at any given point in time. Thus, you will reduce the need to over-provision for individual workload spikes.

In general, virtualization at the data center is applied very broadly – not just to server virtualization. It provides the ability to simulate the availability of hardware that may not be present in sufficient amount – or at all! Virtualization uses the available physical resources as a shared pool to emulate missing physical resources. Virtualization is capable of very fine control over how and to what extent a physical resource is used by a specific virtual machine or server. Thus we have the concept of virtual computer memory (which is not "real" memory but appears to be real) and virtual data storage.

The next sections give details on virtualization technologies at the data center and explain how those technologies are usually the first and most important step we can take in creating energy-efficient and green data centers.

The Concepts of Consolidation and Virtualization

The IT infrastructure energy efficiency strategy consists of centralizing data centers, consolidating IT resources at those data centers, virtualizing the physical IT resources, and integrating applications. Server consolidation and server virtualization both reduce energy use by reducing the number of physical servers, but they use different methods. Server virtualization allows you to keep all of your servers, but they become "virtual servers" when many physical servers share the same physical machine. The diagrams and descriptions of the concepts of consolidation and virtualization were based on the descriptions in the IBM Redpaper "The Green Data Center: Steps for the Journey" (Ebbers et al., 2008). These diagrams and descriptions should clarify the difference and some of the pros and cons of the two methods: consolidation and virtualization.

Consolidation: A Key in Energy Efficiency

A common server consolidation example that I've seen with many projects over the past few years is the consolidation of email servers. As discussed at the beginning of this chapter, for reasons of cost reduction and server management efficiency, there are significant advantages to moving servers to a central data center. As part of the distributed computing architecture where smaller servers were distributed throughout the company, we had email servers that were distributed, often one for each corporate facility with often only a couple hundred users for each server. When the email servers were centralized, dozens of smaller servers could be consolidated onto one or two large mail servers. This was more than consolidating the physical servers onto one large physical server; the large centralized email servers only had one copy of the email application. So server consolidation refers to both consolidating physical servers and consolidating the application.

Figure 3.10 illustrates this idea of consolidation and the energy efficiencies to be gained. Let us assume we have four systems, each running two applications (APP). Also, each machine consumes 2 kW power, 8 kW in total. However, as is often the case for small x86 servers, they are utilized at only 10%. If we are able to consolidate these eight applications to a single, more powerful server and run their operation at a utilization of 70% with a power usage of 4 kW, this single server can operate more energy efficiently. In addition, if we perform a simple power management technique of switching off the previous four systems, the result is a total power consumption of 4 kW and a 70% utilized system.

It's important to note that a decrease in overall power consumption is not the only factor. Hand in hand with the power reduction goes the same amount of heat load reduction and another add-on for the infrastructure. This double reduction is the reason why consolidation is an enormous lever to moving to a green data center.

However, a particular drawback of consolidation is that none of systems 1 through 4 is allowed to be down during the time that the respective applications are moving to the consolidated system. So, during that migration time, higher demands on resources might occur temporarily.

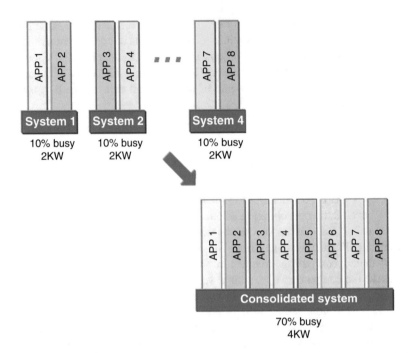

Fig. 3.10 Consolidation of applications from underutilized servers to a single, more efficient server

Virtualization: The Greenest of Technologies

An alternate method to consolidation is virtualization, the concept of dealing with abstract systems. As discussed at the beginning of this chapter, virtualization allows consolidation of physical servers without requiring application consolidation. So, as discussed earlier, with server virtualization we can take ten servers with completely different applications and consolidate them onto one large physical server, where each of the ten stand-alone servers can retain their server name, IP address, etc. The virtual servers still look to users as if they are separate physical servers, but through virtualization, we can dramatically reduce the amount of IT equipment needed in a data center.

Virtualization eliminates the physical bonds that applications have to servers, storage, or networking equipment. A dedicated server for each application is inefficient and results in low utilization. Virtualization enables "carpooling" of applications on servers. The physical car (server) may be fixed, but the riders (applications) can change, be very diverse (size and type), and come and go as needed.

The example in Fig. 3.10 shows how specific applications were moved to another system with a better energy footprint. In the simple case illustrated, we assume all systems are running at the same operating system level. However, what if the applications require different operating system levels or even completely different operating systems? That is where virtualization comes into play.

The term virtualization is widely used and has several definitions, as follows it:

- Can create logical instances of a computer system consisting of CPU, memory, and I/O capabilities
- Can be put together from other virtual components
- Can consist of a virtual CPU or virtual memory and disk
- Can be a virtual network between a virtual computer and the outside world

To have real work done by a virtual system, the virtual system must run on a real system. Obviously additional intelligence is required to do this. There are pure software solutions, or a system's firmware may offer virtualization features, or such features may be hardwired into the system. Many of the current processor architectures have virtualization features integrated, which can be taken advantage of by software solutions such as the IBM System z and p machines. In the field, various other solutions are available, such as VMware Server, VMware ESX, Microsoft Virtual Server, and Xen.

To continue with our example, using virtualization gives a slightly different picture, as shown in Fig. 3.11. Instead of moving the applications to the consolidated server, we now virtualize the existing systems 1 through 4 on our consolidation target. The effect is clear: not only is the application moving, but its complete operating environment has moved with it. Taking a closer look, we find other attractive features:

- Consider the three separate systems. To communicate, they require a network infrastructure such as NICs, cables, and switches. If our virtualization system supports network virtualization, this infrastructure is no longer needed. The virtualized systems can communicate using the virtualization system's capabilities, often transferring in-memory data at enormous speed. Performance and energy efficiency increase because the network components are dropped. Once again, this method reduces the need for site and facilities resources.

- Each of the separate systems has its own storage system, namely, disks. The virtualized systems can now share the disks available to the virtualization system. By virtualizing its storage, the virtualization system can provide optimal disk capacity – in terms of energy efficiency – to the virtualized systems.

Server Virtualization

This section discusses the techniques that are available for server virtualization, the most attractive approach to consolidation. In many cases, it is the easiest and most effective way to transfer workload from inefficient, underutilized systems to efficient, well-utilized equipment.

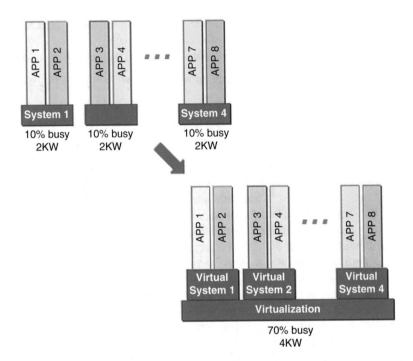

Fig. 3.11 Virtualization allows us to consolidate systems and keep the same server names, etc.

Partitioning

Partitioning is sometimes confused with virtualization, but the partitioning feature is really a tool that supports virtualization. Partitioning is the ability of a computer system to connect its pool of resources (CPU, memory, and I/O) together to form a single instance of a working computer or logical partition (LPAR). Many of these LPARs can be defined on a single machine, as long as resources are available. Of course, other restrictions apply, such as the total number of LPARs a machine can support. The power supplied to the existing physical computer system is now used for all these logical systems, yet these logical systems operate completely independently from one another. LPARs have been available on the IBM System z since the late 1980s and on System p since approximately 2000. Although the System z and System p partitioning features differ in their technical implementations, both provide a way to divide up a physical system into several independent logical systems.

Other Virtualization Techniques

Many virtualization techniques are available, in addition to partitioning. Popular in the market are the VMware products, Xen, and Microsoft Virtual Server. Also, hardware manufacturers extend their products to support virtualization.

VMware ESX Server and Microsoft Virtual Server come with a hypervisor that is transparent to the virtual machine's operating system. These products fall into the full virtualization category. Their advantage is their transparency to the virtualized system. An application stack bound to a certain operating system can easily be virtualized, as long as the operating system is supported by the product.

VMware offers a technology for moving servers called VMotion. By completely virtualizing servers, storage, and networking, an entire running virtual machine can be moved instantaneously from one server to another. VMware's VMFS cluster file system allows both the source and the target server to access the virtual machine files concurrently. The memory and execution state of a virtual machine can then be transmitted over a high-speed network. The network is also virtualized by VMware ESX, so the virtual machine retains its network identity and connections, ensuring a seamless migration process. System p Live Partition Mobility offers a similar concept.

Xen uses either the paravirtualization approach (as the POWER architecture does) or full virtualization. In the partial approach (paravirtualization), virtualized operating systems should be virtual-aware. Xen, for example, requires virtual Linux systems to run a modified Linux kernel. Such an approach establishes restrictions to the usable operating systems. However, while they are hypervisor-aware, different operating systems with their application stacks can be active on one machine. In the full approach, the hardware, such as Intel's Vanderpool or AMD's Pacifica technology, must be virtual-aware. In this case, running unmodified guests on top of the Xen hypervisor is possible, gaining the speed of the hardware.

Another technique is operating system level virtualization. One operating system on a machine is capable of making virtual instances of itself available as a virtual system. Solaris containers (or zones) are an example of this technique. In contrast to the other techniques, all virtualized systems are running on the same operating system level, which is the only operating system the machine provides. This can become a very limiting restriction, especially when consolidating different server generations. Often the application stack is heavily dependent on the particular operating system. We reach a dead end when we want to consolidate servers running different operating systems such as Windows and Linux.

Chapter Summary and Conclusions

The following conclusions can be made from the above discussion on virtualization and green IT:

- Virtualization is the most promising technology to address both the issues of IT resource utilization and facilities space, power, and cooling utilization.
- Many healthcare facilities are addressing the situation from end to end, at the server end through power management features and at the data-center ends through integrated IT/facilities modular solutions.
- IT virtualization includes server virtualization, storage virtualization, client virtualization, virtualization using cluster architecture, and virtualization of blade servers.
- The ultimate objective and benefit of virtualization is the significant IT flexibility it brings healthcare facility users. IT virtualization can benefit data protection, business continuity, and disaster recovery.

Additional Reading

The future of mobile E-health application development: Exploring HTML5 for context-aware diabetes monitoring, The 3rd international conference on current and future trends of information and communication technologies in healthcare (ICTH-2013), *Procedia Computer Science* 21 (2013) 351–359.

Kilobyte, megabyte, gigabyte, terabyte, petabyte, exabyte, zettabyte, defined at http://www.taringa.net/post/info/1196306/Kilobyte-Megabyte-Gigabyte-Terabyte-Petabyte-Exabyte-Z.html. Accessed on 23rd Oct 2016.

Memory sizes explained – gigabytes, terabytes & petabytes in Layman's Terms – http://www.makeuseof.com/tag/memory-sizes-gigabytes-terabytes-petabytes/. Accessed on 23rd Oct 2016.

Health care in India – vision 2020 issues and prospects. Srinivisan. Accessed on 23rd Oct 2016 at the URL mentioned below: http://planningcommission.gov.in/reports/genrep/bkpap2020/26_bg2020.pdf.

Mobile based primary health care system for rural India, M V Ramana Murthy, Mobile Computing and Wireless Networks, CDAC, Electronics city. Accessed on 23rd Oct 2016 at the URL – https://www.w3.org/2008/02/MS4D_WS/papers/cdac-mobile-healthcare-paper.pdf.

Bio-IT and Healthcare in India, the Joint Report (March 2014) NASSCOM and department of Biotechnology Ministry of Science and Technology Government Of India. Accessed on 23rd Oct 2016 at the URL http://dbtindia.nic.in/wp-content/uploads/able_bio_it_report_final_light.pdf.

What is telemedicine? – Accessed on 27th Oct. 2016 at the URL – https://www.icucare.com/PageFiles/Telemedicine.pdf.

Telemedicine: patient privacy rights of electronic medical records by Chari J. Young accessed on 27th Oct 2016 at the URL – http://heinonline.org/HOL/LandingPage?handle=hein.journals/umkc66&div=45&id=&page=.

The contents of digital medical records can be seen at the URL mentioned below Accessed on 2nd Nov 2016 – http://www.medicalrecords.com/what-is-an-electronic-medical-record.

Telemedicine Journal and e-Health by Michael Ackerman, Frank Ferrante, Mary Kratz, Salah Mandil. Accessed 8th May 2017 at http://online.liebertpub.com/doi/abs/10.1089/15305620252933419?journalCode=tmj.2.

The benefits of mobile health strategies. Accessed 8th May 2017 at http://mhealth-intelligence.com/news/the-benefits-of-mobile-health-strategies.

Technology is completely revolutionizing the healthcare industry, the blog by Tom Raftery. Accessed on 9th May 2017 at http://greenmonk.net/2014/05/28/technology-is-completely-revolutionising-the-healthcare-industry/.

12 Elements to support a viable Health IT and Telehealth infrastructure by Bob Herman. Accessed 9th May 2017 at http://www.beckershospitalreview.com/healthcare-information-technology/12-elements-to-support-a-viable-health-it-and-telehealth-infrastructure.html.

Telemedicine: Challenges and opportunities, an editorial article by Anthony C. Smith cited at http://www.tandfonline.com/doi/pdf/10.1586/17434440.4.1.5? needAccess=true. Retrieved 20th May 2017.

Telemedicine: Challenges and opportunities, published in Journal of High Speed Networks 9(1):15–30 January 2000 and cited at https://www.researchgate.net/publication/220200890_Telemedicine_Challenges_and_opportunities. Retrieved 20th May 2017.

Adoption of Telemedicine – challenges and opportunities, research paper by Ramakrishna Dantu and RadhaKanta Mahapatra from Information Systems and Operations Management College of Business University of Texas at Arlington, cited at https://pdfs.semanticscholar.org/8267/cb4b468233850adfe1ea-b1387ef2d1202e5d.pdf retrieved on 20th May 2017.

Ghosh, R., & Ahadome, T. (2012). Telehealth's promising global future. Retrieved 3 Feb 2013, from Analytics: http://www.analytics-magazine.org/januaryfebruary-2012/508-telehealths-promising-global-future.

Opportunities and challenges for telemedicine, an article published in Health Capital, April 20111, Volume 4, Issue 4. Retrieved on 20th May 2017 at https://www.healthcapital.com/hcc/newsletter/04_11/tele.pdf.

PowerPoint presentations relating to Telemedicine can be accessed at the URLs quoted below, all accessed on 20th May 2017

https://www.slideshare.net/khandhar/telemedicine-ppt.
https://www.slideshare.net/DelightsonRufus/telemedicine-32936573.
https://www.slideshare.net/rupak/telemedicine-presentation-569817.
https://www.slideshare.net/darshilshah940098/telemedicine-53764406.
https://www.slideshare.net/HowardRcis/telemedicine-presentation-feb-2014.
https://www.slideshare.net/alternative9595/
    telemedicine-presentation-32138394.
https://www.slideshare.net/careclix/all-about-telemedicine.
https://www.slideshare.net/AdityaMaduriya/telemedicine-63767663.
https://www.slideshare.net/bamit/
    telemedicine-an-opportunity-in-healthcare-in-india.
https://www.slideshare.net/AkritiSingh9/telemedicine-51791745.

The challenges and opportunities of telemedicine by Tom Rodgers published November 18, 2015, retrieved http://www.mckesson.com/blog/the-challenges-and-opportunities-of-telemedicine/.

Chapter 4
Using Data Science to Make Healthcare Green

The 4 big reasons why healthcare needs data science

— IDG report 2015

This chapter describes significant ways that data science (Provost & Fawcett, 2013) can be used to make healthcare green (Godbole & Lamb, 2015). Here are the four big reasons why healthcare needs data science from a 2015 IDG report.

Reason 1: Hospital Claims Data

In 2010, there were 35.1 million discharges in the United States, with an average length of stay of 4.8 days according to the National Hospital Discharge Survey. That same survey went on to note that there were 51.4 million procedures performed. The National Hospital Ambulatory Medical Care Survey in 2011 stated that the number of outpatient department visits was 125.7 million with 136.3 million emergency department visits. These are some of the basic figures showing the amount of care the US healthcare system has provided. Using data science to annualize this sort of data allows healthcare providers to start building a new intuition built on a data narrative that could possibly help avoid the spread of diseases or address specific health threats. Using a combination of descriptive statistics, exploratory data analysis, and predictive analytics, it becomes relatively easy to identify the most cost-effective treatments for specific ailments and allows for a process to help reduce the number of duplicate or unnecessary treatments. The power in predicting a future state is in using that knowledge to change the behavior patterns of today.

Reason 2: Clinical Data

This sort of data takes the form of doctor's notes, lab results, and medical images gathered during a patient's encounter with a healthcare provider. For example, it is routine for hospitals today to use natural language processing algorithms to analyze patient records so they may identify certain individuals at risk for medical conditions. Recently it was reported that healthcare providers failed to recognize three high blood pressure readings at separate visits in 26% of pediatric patients reviewed by the American Medical Association. Recognizing these types of patterns in the face of growing data will only become more difficult over time for healthcare providers.

© Springer International Publishing AG, part of Springer Nature 2018
N. S. Godbole, J. P. Lamb, *Making Healthcare Green*,
https://doi.org/10.1007/978-3-319-79069-5_4

Reason 3: Pharma R&D Data

In recent years, there have been a number of partnerships developed between pharmaceutical companies. Consider the Project Data Sphere, an initiative to share, integrate, and analyze historical cancer trial data sets for the purpose of accumulating research findings and accelerating cures. The power of this rich dataset is in the analysis and the global focus on finding solutions for cancer patients.

Reason 4: Patient Behavior and Sentiment Data

A study by AMI Research suggests that "wearables" are expected to reach $52 million by 2019. Wearables monitor heart rates, sleep patterns, walking, and much more while providing new dimensions of context, geolocation, behavioral pattern, and biometrics. Combine this with the unstructured "lifestyle" data that comes across social media, and you have a potent combination that is more than just numbers and tweets.

There is no doubt that data science is becoming more and more significant for healthcare. It is reasonable to expect that we will likely recover more quickly from illness and injury, live longer because of newly discovered drugs, and benefit from more efficient hospital surgeries – and in large part this will be because of how we analyze Big Data. As healthcare continues to fully embrace data science, it will change the future for everyone.

There are also the aspects of saving money using energy-efficient data science. A typical data center of 25,000 ft^2 with electrical costs of 14 cents per KWH will cost about $2.5 million a year in electricity. Going green can easily cut that yearly electrical cost in half. Data science applied to going green is especially relevant when looking at ways to improve software and application efficiency. The following sections give examples.

Implementing Efficient Applications and Deduplicating Data

Software and application efficiency can be very significant for green healthcare and green IT. The author has had recent experience where the procedure for creating a data warehouse report was reduced from 8 h to 8 min merely by changing the Oracle data warehouse search procedure (e.g., don't search the entire database each time when only a much smaller search is required). During the 8 h required to create the report, the large server was running at near peak capacity. Sure, that type of significant application inefficiency has been created and fixed many times over the history of programming. But what about the cases where a few application efficiencies will make an application run 20% faster? That 20% more efficient application will also result in 20% lower energy use. The steps required to improve application efficiency by a few percent are often not easy to determine. However, the added incentive of saving energy, while making the application run faster, is a significant plus.

Data-storage efficiency, such as the use of tiered storage, is also very significant. Data deduplication (often called "intelligent compression" or "single-instance storage") is a method of reducing storage needs by eliminating redundant data. Only one unique instance of the datum is actually retained on storage media, such as disk or tape. Redundant data are replaced with a pointer to the unique data copy. For example, a typical email system may contain 100 instances of the same 1-megabyte (MB) file attachment. If the email platform is backed up or archived, all 100 instances are saved, requiring 100 MB storage space. With data deduplication, only one instance of the attachment is actually stored; each subsequent instance is just referenced back to the single saved copy. In this example, a 100 MB storage demand can be reduced to only 1 MB.

Data deduplication offers other benefits. Lower storage space requirements will save money on disk expenditures. The more efficient use of disk space also allows for longer disk-retention periods, which provides better recovery time objectives (RTO) for a longer time and reduces the need for tape backups. Data deduplication also reduces the data that must be sent across a WAN for remote backups, replication, and disaster recovery.

Data deduplication uses algorithms to dramatically reduce the amount of storage space needed. Many organizations are dealing with increased scrutiny of electronically stored information because of various regulations; this need to preserve records is driving significant growth in demand for storing large sets of data. Depending on the type of information being compressed, deduplication can enable a compression rate of between 3:1 and 10:1, allowing businesses to reduce their need for additional storage equipment and associated tapes and disks. Many businesses are already using the technology.

Using Data Mining to Establish the Applicable New Technologies to Significantly Make Health Green

This section describes some of the emerging technologies that could significantly improve green IT at hospitals.

Fuel Cells for Data-Center Electricity

Fuel cells have been proposed to power data centers. For instance, the polluting diesel backup generators that most data centers rely on may be replaced by fuel cells. Back in 2008, Fujitsu began using a fuel-cell generator to power its data center in Silicon Valley. Fuel cells have also been proposed to be used in an emergency or during peak demand to take some of the load off the grid. Hydrogen-powered fuel cells are very environmentally desirable since the only output, in addition to energy, is water. The problem is in obtaining the hydrogen. Currently hydrogen is usually produced through a very energy-intensive process using natural gas and immense amounts of electricity. When technological breakthroughs allow us to produce hydrogen efficiently, then fuel cells for data-center energy will be a significant step forward.

Other Emerging Green Technologies for the Data Center

Energy costs will likely continue to rise in the future as will the computing requirements of most organizations. Taking steps today to increase the efficiency of the cooling system can offset the impact of rising energy costs when newer, higher-efficiency technologies are deployed. Three technologies, in particular, have potential to significantly enhance data-center energy efficiency:

- Multicore processors
- Embedded cooling
- Chip-level cooling
- SSD (solid-state device) storage technology

Newer servers are now based on multicore processors that enable a single processor to perform multiple separate tasks simultaneously, run multiple applications on a single processor, or complete more tasks in a shorter amount of time. Chip manufacturers claim that multicore processors can reduce power and heat by up to 40% (Lamb, 2009).

Embedded cooling uses the cooling infrastructure to deliver high-efficiency cooling directly inside the rack. This approach brings cooling even closer to the source of heat and allows the cooling system to be optimized for a particular rack environment. This type of system is able to prevent heat from the system from entering the room by removing the heat before it leaves the rack.

Chip-level cooling takes this approach to the next level by helping to move heat away from the chip. As embedded and chip-level cooling solutions are deployed, a highly efficient three-tiered approach to data-center cooling will emerge. In this approach, heat is effectively moved away from the chip and then cooled in the rack, with stable temperatures and humidity maintained by room air conditioners. These developments are not expected to reduce data-center cooling requirements. Instead, they will result in an increase in the amount of computing power that can be supported by a particular facility. As a result, the efficiency improvements made today will continue to pay dividends well into the future as these new developments enable existing facilities to support densities that are not possible today.

The cooling system represents a significant opportunity to improve efficiency. In many cases, relatively simple and inexpensive changes such as improving room sealing, moving cables, or other objects that obstruct airflow or installing blanking panels can pay immediate dividends. In addition, new technologies, such as variable capacity room air conditioners and sophisticated control systems, should be considered for their impact on efficiency. Finally, supplemental cooling systems provide a response to increased equipment densities that can increase the scalability and efficiency of existing cooling systems.

SSD (solid-state device) storage technology can significantly reduce energy needed for storage and significantly increase storage access speed. SSD is the same technology used in our flash drives, but for data centers SSD is used to replace terabytes of spinning disk technology. In the author's experience, SSD can greatly improve the speed for large database applications since data access is immediate due to random access and the database program doesn't experience the delays incurred waiting for the spinning disk to reach the start of the data needed. As we know from the use of our flash drives, SSD technology requires very little power and requires very little cooling. The cost of SSD continues to come down, and we can expect spinning disk technology to continue to be replaced by SSD. That's great news for energy efficiency at data centers.

Adopting Data Analytics to Help Make Healthcare Green

A good way to use data analytics (Dietrich, Plachy, & Norton, 2014) to help make healthcare green is to use the tried and true Six Sigma process (Arthur, 2011).

Six Sigma is a set of techniques and tools for process improvement. Their techniques, originally developed by Motorola in the 1980s, are being used by many industries throughout the world. The Six Sigma steps applicable for Green IT and sustainability (Johnson & Johnson, 2012), (Franchetti, 2013) are:

1. Define
2. Measure
3. Analyze
4 Improve
5. Control

These five steps are similar to the five-step process that IBM has used for several years for creating energy-efficient "green" data centers: (1) diagnose, (2) manage and measure, (3) use energy-efficient cooling, (4) virtualize, and (5) build new or upgrade facilities when feasible. These Six Sigma steps for green IT were first mentioned in Chap. 2 and are also discussed at the beginning of Chap. 6. The authors believe these five steps are key to helping make healthcare green, and the Six Sigma steps are indicated in the case studies discussed in Chaps. 10 and 11.

Six Sigma Approach to Help Make Healthcare Green

Define the Opportunities and Problems (Diagnose)

The step here is to do a data-center energy-efficient assessment. The assessment should include a list of unused IT equipment that can be turned off. In addition, the diagnostic phase can help encourage organizations to retire unused software applications and focus on adopting more effective software that requires fewer CPU cycles. A typical x86 server consumes between 30% and 40% of its maximum power when idle. IT organizations should turn off servers that don't appear to be performing tasks. If anyone complains, organizations should look into whether the little-used application can be virtualized. Check with your electric utility. Some utilities offer free energy audits.

Measure (You Can't Manage What You Can't Measure)

Many hardware products have built-in power management features that are never used. Most major vendors have been implementing such features for quite some time. These features include the ability of the CPU to optimize power by dynamically switching among multiple performance states. The CPU will drop its input voltage and frequency based on how many instructions are being run on the chip itself. These types of features can save organizations up to 20% on server power consumption.

Analyze IT Infrastructure with Energy-Efficient Cooling

Many data centers may use hot aisle/cold aisle configurations to improve cooling efficiency, but there are also some small adjustments they can make. Simple "blanking panels" can be installed in server racks that have empty slots. That's a great way to make sure the cold air in the cold aisle doesn't start mixing with the hot air in the hot aisle any sooner than it needs to. Organizations should also seal cable cutouts to minimize airflow bypasses. Data organizations should consider air handlers and chillers that use efficient technologies such as variable frequency drives which adjust how fast the air conditioning system's motors run when cooling needs dip.

Improve (Virtualize IT Devices – Use Cloud When Feasible)

Virtualization continues to be one of the hottest green data-center topics. Many current server CPU utilization rates typically hover between 5% and 15%. Direct-attached storage utilization sits between 20% and 40%, with network storage between 60% and 80%. Virtualization can increase hardware utilization by 5–20 times and allows organizations to reduce the number of power-consuming servers. Cloud computing is the "ultimate" in virtualization, and we'll discuss virtualization and cloud computing in more detail later in this book.

Control (Continue to Measure and Manage and Upgrade as Necessary)

Going green is easiest if you are building a new data center. First, you make a calculation of your compute requirements for the foreseeable future. Next, you plan a data center for modularity in both its IT elements and its power and cooling. Then you use data-center modeling and thermal assessment tools and software – available from vendors such as APC, IBM, and HP – to design the data center. The next step is to procure green from the beginning – which partly means buy the latest equipment and technologies such as blade servers and virtualization.

Once you have the equipment, you integrate it into high-density modular compute racks, virtualize servers and storage, put in consolidated power supply, choose from a range of modern cooling solutions, and, finally, run, monitor, and manage the data-center dynamics using sensors that feed real-time compute, power, and cooling data into modern single-view management software that dynamically allocates resources.

Chapter Summary and Conclusions

In this chapter, we looked at the different ways data science can be used to help determine the best ways to make healthcare green. The Six Sigma approach to making healthcare green is a well-proven technique. As discussed in this chapter, data science and Big Data analytics are excellent ways to make healthcare green. Those techniques combined with emerging technologies such as mobile computing and cloud computing hold great promise to reduce costs and significantly improve services in the healthcare industry. The continued significant use of electric energy for the IT infrastructure used to support the healthcare industry has increased pressure on the industry to support green IT initiatives and overall sustainability. Healthcare green IT efficiency improvements must be made in compliance with the expanding regulations to protect patient privacy (Godbole, 2009).

There is an ever-growing amount of excellent data (Big Data) for the healthcare sector, and the use of Big Data analytics along with data science in general will lead to ever-improving results. Going forward, we in IT all have a role in helping improve the outlook for healthcare by contributing to IT infrastructure electric energy sustainability, data protection, and the continued improvement in cloud computing for IT cost reduction along with improved data protection.

Additional Reading

Big Data Analytics for Healthcare. The tutorial presentation at the SIAM International Conference on Data Mining Austin, TX, 2013 by Jimeng Sun from Healthcare Analytics Department, IBM TJ Watson Research Center and Chandan K. Reddy from Department of Computer Science, Wayne State University. For the updated tutorial slides – visit the link http://dmkd.cs.wayne.edu/TUTORIAL/Healthcare/.

Using Mobile & Personal Sensing Technologies to Support Health Behavior Change in Everyday Life: Lessons Learned. By PredragKlasnja, Sunny Consolvo , David W. McDonald, James A. Landay & Wanda Pratt, University of Washington, Seattle, WA – the paper available at the URL mentioned below, accessed on 27th October 2016. http://citeseerx.ist.psu.edu/viewdoc/download?doi=10.1.1.481.4 82&rep=rep1&type=pdf.

Emerging Technologies for Chronic Disease Care. By Juliann Schaeffer accessed on 27th October 2016 at the URL: http://www.fortherecordmag.com/archives/073012p10.shtml.

mHealth technologies for chronic disease prevention and management, December 2015 Report. By Sax Institute L Laranjo, A Lau, B Oldenburg, E Gabarron, A O'Neill, S Chan, E Coiera, accessed on 27th October 2016 at the URL: http://www.saxinstitute.org.au/wp-content/uploads/mHealth-Technologies-for-Chronic-Disease-Prevention-and-Management-1.pdf.

Accelerating Commercialization of Cost-Saving Health Technologies. Accessed on 27th October 2016 at the URL: http://jacobsschool.ucsd.edu/vonliebig/docs/Accelerating_commerce.pdf.

The Diffusion of New Technology: Costs and Benefits to Health Care. PETER J. NEUMANN and MILTON C. WEINSTEIN – accessed on 27th October 2016 at the URL mentioned below: https://www.ncbi.nlm.nih.gov/books/NBK234309/.

The Effectiveness of Mobile-Health Technology-Based to Disease Management Interventions for Health Care Consumers by OmniTech. Accessed on 27th October 2016 at the URL: https://novoed.com/mhealth/reports/52124.

Chapter 5
Green IT, Cloud Technologies, and Carbon Footprint in Hospitals

Knowing is not enough; we must apply. Willing is not enough; we must do.

—Goethe

Sustainability for the healthcare sector needs to be addressed from both energy and carbon-footprint aspects. The healthcare sector is complex. The authors believe that green IT is an ideal way for hospitals to make a significant step in the green direction for several reasons. First, IT is continually being refreshed at hospitals and most organizations as part of the need for the hospitals to keep vital technology up to date. That refresh cycle is similar to the refresh cycle for our company-owned laptops, i.e., every 3 or 4 years. This results in a tremendous amount of eWaste, i.e., electronic waste generated in hospitals, and yet this is a neglected area. At all hospitals, servers and data storage are continually being replaced. Replacing the IT equipment and upgrading the application architecture with energy-efficient systems such as virtual servers, virtual data storage, and efficient application and database structures can easily reduce IT power consumption for the replaced equipment by 50%. A second compelling reason to move to green IT is that virtual server and virtual data-storage technology are methods that allow hospitals to reduce equipment and system management costs. Include private cloud computing for both your production and test/development systems, and the savings are even greater. So the latest green technology is based around a very solid business case without even considering the savings due to reduction in energy costs.

This chapter will describe a process for creating green IT at a hospital based on the extensive work done for green IT in other sectors (mainly data centers) in effectively measuring and improving IT energy efficiency.

© Springer International Publishing AG, part of Springer Nature 2018
N. S. Godbole, J. P. Lamb, *Making Healthcare Green*,
https://doi.org/10.1007/978-3-319-79069-5_5

Calculating a Hospital's IT Energy Efficiency and Determining Cost-Effective Ways for Improvement

The environmental impact of the healthcare sector has become an important factor globally and is continuing to draw the attention of regulators. The energy consumption of the healthcare sector (whose largest sub-segment is hospitals) has been growing due to many factors. These factors include the rapid growth and adoption of information and communication technology (ICT) in healthcare. The new IT technologies and applications used in healthcare include "cloud computing" and "mMedicine," i.e., "mobility in health," eHealth, and tele(health) care for remote delivery of healthcare services. In general, the healthcare industry needs to reap the benefits of emerging technologies such as mobile computing and cloud computing, along with the use of health information technology (HIT) to help solve the ever-growing operating cost problems. One challenge facing the healthcare sector is how best to calculate the energy efficiency of this very complex sector. The work done over the past few years to analyze and create very energy-efficient data centers presents an excellent opportunity for cost-effective green IT at hospitals.

Green IT is an ideal way for hospitals to make a significant step in the green direction for several reasons. First, IT is continually being refreshed at hospitals and most organizations as part of the need for the hospitals to keep vital technology up to date. That refresh cycle is similar to the refresh cycle for our company-owned laptops, i.e., every 3 or 4 years. Any electronic waste (eWaste) from this refresh cycle must be recycled as part of a sustainability program. Replacing the IT equipment and upgrading the application architecture with energy-efficient systems such as virtual servers, virtual data storage, and efficient application and database structures can easily reduce IT power consumption for the replaced equipment by 50%. A second compelling reason to move to green IT is that virtual server and virtual data-storage technology are methods that allow hospitals to reduce equipment and system management costs. Include private cloud computing for both your production and test/development systems, and the savings are even greater. So the latest green technology is based around a very solid business case without even considering the savings due to reduction in energy costs.

This chapter describes a process for calculating a hospital's IT energy efficiency along with information on the best ways to improve energy efficiency in a cost-effective manner. The process builds on the work done in other sectors (mainly data centers) in effectively measuring and improving IT energy efficiency.

Introduction

Globally, the healthcare sector (Godbole & Lamb, 2013) is emerging in importance and plays a key role in our economy. Healthcare is not just another service industry; the healthcare sector is characterized by the fact that it brings together people and institutions involved in providing healthcare to those who need such services in times of need and stress. Healthcare is not only a complex industry but also a fragmented industry; it is composed of a number of sub-segments such as hospitals (a major sub-segment), pharmaceuticals, medical insurance agencies, medical equipment manufacturers, dealers and distributors, and diagnostics. Of these sub-segments, hospitals, by their very nature of operations, collect PHI (protected health information) (Protected Healthcare Information (PHI), (n.d.))-related entities as well as treatment-specific confidential data.

The healthcare sector makes heavy use of technology including social media (KPMG, 2011 October). Technological advances have brought about revolution in the healthcare industry worldwide from modern testing techniques to improved surgical equipment, remote health monitoring technologies with the help of modern digital equipment, etc. There are a number of online healthcare portals.

The complexity and scale of the healthcare industry helps account for the large environmental (carbon) footprint (UC Hospitals, 2009). For example, in the United States, it is estimated that healthcare facilities generate more than 5.9 million tons of waste annually. Hospitals contribute approximately 8% of the greenhouse gas emissions resulting from human activity. "Sustainability" in the healthcare sector can be classified into two types – (i) sustainability issues that are from non-IT areas and (ii) sustainability issues that are due to practices in the IT infrastructure of healthcare, mainly in hospitals. Of late, sustainability is getting integrated into healthcare investors' financial strategies. Sustainable thinking for healthcare (mainly for hospitals) comes from the "green-triad" – (a) green infrastructure, (b) green IT, and (c) green medical technology:

- Cost savings through use of optimized and modernized heating, ventilation, and cooling (green infrastructure)
- Use of optimized illumination control systems for greater comfort and energy savings (green infrastructure)
- Deployment of customized solutions, services, products, and technologies to optimize the life-cycle performance for hospitals with a view to achieve maximum energy savings, performance, and sustainability without losing safety and comfort (green infrastructure)
- Reduced traveling by using modern communications systems, e.g., videoconferencing, telemedicine, etc. (green infrastructure)
- Green transformation of healthcare data centers (Lamb, 2009) through use of consolidation and virtualization technologies (green IT)
- Improving data-center energy efficiency (Lamb, 2009) – file server storage reduction and usage of groundwater for air cooling in DC (green IT)
- Green ICT architectures – desktop and server virtualization (green IT)
- Energy-efficient medical technologies (green medical technologies)
- Making operation of medical technologies environmentally friendly (green medical technologies)
- Healthcare waste and eWaste (Godbole, 2011a, 2011b) (electronic and electrical equipment) disposal through reduce-reuse-recycle principle (green infrastructure)

Table 5.1 Hospital: functional area-wise typical energy consumption and carbon emissions

Functional area of use	Relative energy consumption (kWh)	Carbon emissions (kilograms CO2-e)
Ward	1350	293
Surgery areas	844	268
Back area	619	176
Consulting areas (consulting rooms, etc.)	510	166
Administration and office blocks	474	154
Corridors (24 h)	349	95

The carbon emissions were calculated using the following Scope 3 emission factors: 1 kWh electricity = 1.35 kg CO2-e and 1 GJ of gas = 55.7 kg CO2-e

Much of the focus today is on the greening of data centers. Although healthcare organizations use a variety of IT applications and infrastructures, patients may not always see that reflected into the costs to the patients as IT is not one of healthcare organizations' primary activities. Moreover, due to rapid obsolescence of information technology, the IT systems and applications used in healthcare need to be updated. The cost of IT systems in healthcare services is quite high, and often, many of healthcare organizations pass this cost to their patients. This is how the healthcare delivery becomes expensive from an end-user perspective.

Sustainability is a major aspect of healthcare sector operations worldwide, in view of the impact of CFCs (chlorofluorocarbon) on our environment and carbon-footprint share of the healthcare. Table 5.1 shows hospital functional areas in terms of their typical relative monthly energy use and carbon emissions.

There have been a number of studies on green practices in the healthcare sector (Godbole, 2011a, 2011b) and barriers to green practices (Muduli, 2012) in the healthcare waste sector. There is indeed a need to rethink healthcare systems. To design more sustainable health systems, advantage must be taken of demand-side opportunities.

Green IT and Carbon Footprint for Hospitals

As indicated above, sustainability for the healthcare sector needs to be addressed from both energy and carbon-footprint aspects. The healthcare sector is complex. The authors believe that green IT is an ideal way for hospitals to make a significant step in the green direction for several reasons. First, IT is continually being refreshed at hospitals and most organizations as part of the need for the hospitals to keep vital technology up to date. This results in a tremendous amount of "eWaste," i.e., electronic waste generated in hospitals, and yet this is a neglected area. At all hospitals, servers and data storage are continually being replaced. Replacing the IT equipment and upgrading the application architecture with energy-efficient systems such as

virtual servers, virtual data storage, and efficient application and database structures, can easily reduce IT power consumption for the replaced equipment by 50%.
A second compelling reason to move to green IT is that virtual server and virtual
data-storage technology are methods that allow hospitals to reduce equipment and
system management costs. Include private cloud computing for both your production and test/development systems, and the savings are even greater. So the latest
green technology is based around a very solid business case without even considering the savings due to reduction in energy costs.

This section describes a process for creating green IT at a hospital based on the
extensive work done for green IT in other sectors (mainly data centers) in effectively measuring and improving IT energy efficiency. The following six tasks are
applicable to all green IT projects, including hospitals.

Communicate Green IT Plans and Appoint an Energy Czar

Being able to measure the current state of affairs, energy-wise, is one of the first
steps to take. You need a baseline on which to start measuring the impact of your
hospital's energy-saving initiatives in the green IT area. Of course, you must also
communicate your proposed energy efficiency initiatives right away. Let all hospital
employees know about the plans and goals to save energy via green IT. Besides
communicating with your hospital employees, set up an organization to drive the
effort. You may start by making one person responsible; give that person a title
("energy czar").

Consolidate and Virtualize

Consolidating IT operations and using virtualization to reduce server footprint and
energy use are the most well-recognized and most-often-implemented efficiency
strategies of the past few years. Most of the largest technology organizations in the
world have completed major data-center consolidation projects. The projects
included server consolidation and virtualization.

Install Energy-Efficient Cooling Units

Cooling can account for almost half of your IT electrical energy use. State-of-the-
art cooling technology can significantly reduce cooling costs.

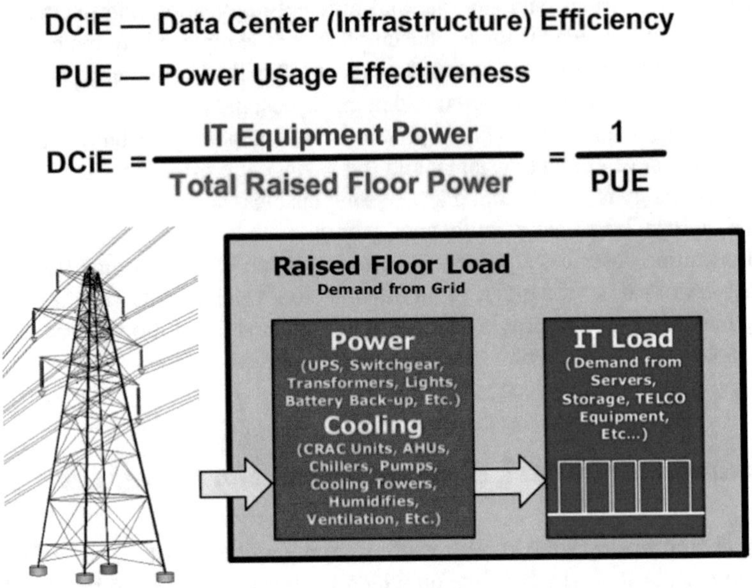

DCiE — Data Center (Infrastructure) Efficiency

PUE — Power Usage Effectiveness

$$DCiE = \frac{\text{IT Equipment Power}}{\text{Total Raised Floor Power}} = \frac{1}{\text{PUE}}$$

Fig. 5.1 Data-center metrics DCiE and PUE (Source: IBM Green Data Center)

Measure and Optimize

The metrics used to measure green IT for data centers are the power usage effectiveness (PUE) and the related Data Center Infrastructure Efficiency (DCiE).

In 2008, the Green Grid modified the PUE metric to make it simpler to understand. The new metric is called Data Center Infrastructure Effectiveness (DCiE). The following graphic describes both metrics:

$$PUE = \frac{\text{Meter total power}}{\text{UPS output power}}$$

A PUE below 1.5 is considered excellent (a PUE of 1.0 is "perfect") (Fig. 5.1).

The fast way often used to compute PUE at a data center is to use information from the uninterruptable power supply (UPS). The output of the UPS is the power used for IT. The meter reading for the data center gives the total power needed.

Make Use of Rebates and Incentives

More utility providers are offering rebates or other incentives that encourage businesses to update equipment and adopt efficient operational practices that can help reduce peak and total power demands.

Since power companies don't want to have to build all the new capacity that may be required to keep all these new and growing data centers in operation, everyone has ended up happy. The facilities personnel didn't have to build as much space; the IT organization and engineering groups got new equipment that was smaller, cooler, and faster than before; and the power company was happy to eliminate a big chunk of demand.

IT Infrastructure Architecture in Healthcare Systems

With the rapid growth of healthcare industry, the reliance on IT increases dramatically. Newer strategies need to be used in IT infrastructure for healthcare industry to be able to cater to the need for reducing the cost of healthcare delivery, improving the quality of healthcare delivery, and extending the reach of healthcare services, especially in the areas where use of cloud-computing-based applications has a great potential (Lamb, 2011; Padhy, Patras, & Satapathy, 2012), for example, making medical expertise available to rural people through a collaborative use of cloud-based medical databases of past history and patterns for use of patient treatment. There is also a need for innovative IT solutions to support the (i) ability to survive the inevitable changes in healthcare industry and (ii) to enable a competitive advantage for the industry. Given the rising costs of healthcare service operations and the need to improve the reach of healthcare services, mobile medicine is becoming an imperative. In healthcare IT infrastructure, both the Internet and wireless sensor networks are deployed to achieve mobility in monitoring patient conditions. For example, sensors are attached to patients for monitoring and measuring their vital signs, e.g., patient's heart rate or body temperature. Such measurements are periodically sent to the medical staff. The challenge in IT infrastructure design for the healthcare lies in managing the trade-off between the level of security (Godbole, 2009) provided and the resources consumed. Mobile access can provide significant benefits such as easy access to medical information and energy savings. However, security and privacy issues under mobile access need to be carefully considered.

The healthcare service delivery cost issue was mentioned in an earlier section of the chapter. A cloud-based architecture would be especially appropriate for a rural healthcare center. In that architecture, a cloud controller would manage physical resources, monitor the physical machines, place virtual machines, and allocate storage. The controller would respond to new requests or changes in workload by provisioning new virtual machines and allocating physical resources. Note that a cloud service provider would need to use an authentication and authorization mechanism. "Authentication" means that each user has an identity which can be trusted as genuine. "Authorization" means that each resource will have a set of users and groups that can access it.

Case Studies for Data Centers as Models for Hospitals

This section includes green IT case studies for data centers that carry over to hospitals. The case studies are for data centers from publications by the authors (Lamb, 2009, 2011). The energy savings can be significant. IBM has found that a typical data center of 25,000 ft^2 with electrical costs of 14 cents per KWH will cost about $2.5 million a year in electricity. Going green can easily cut that yearly electrical cost in half.

Employing the Proposed Process: Case Studies from Data Centers

Here are three case studies showing results of reduction of energy use at data centers based on the five-step process described in Chap. 2. These case studies would carry over to hospitals.

- *Bank in South Africa*

  - Speed: Virtual servers and use of private cloud reduced test system setup time from 2 weeks to 2 h.
  - Energy Savings: The bank reported a reduction in virtual servers for test/dev by half, reducing power and cooling in half.

- *Lexington, Kentucky, large data center*

  - Equipment Reduction: Originally this 61,000 ft^2 data center had over 2500 physical servers with 59% of the servers having monthly utilization under 5%. There was no room for additional servers. Server virtualization eventually reduced the amount of equipment needed by 80% with very few servers having low monthly CPU utilization.
  - Energy Savings: Total energy use for this very dynamic data center was reduced significantly, and the data center has continued to add customer and virtual servers.

- *Data Center in India*

  - Equipment Reduction: Consolidation reduced the number of data centers from 35 to 7 with significant server virtualization reducing the total number of servers needed by over 50%.
  - Energy Savings: Total energy use for IT and cooling reduced by over 50%.

Relating Sustainability to Cost in the Healthcare Sector

The next challenge would be finding the link between sustainability and cost of healthcare delivery. As mentioned earlier, IT is not one of a healthcare organization's primary activities. Therefore, we envisage that most stakeholders in healthcare would not consider "greening of IT" as their immediate agenda. However, overall healthcare greenness and cost would definitely be a significant concern.

Chapter Summary and Conclusions

In this chapter, we looked at the role of green IT, cloud technologies, and carbon footprint in making healthcare green. As discussed in this chapter, emerging technologies such as mobile computing and cloud computing hold great promise to reduce costs and significantly improve services in the healthcare industry. The continued significant use of electric energy for the IT infrastructure used to support the healthcare industry has increased pressure on the industry to support green IT initiatives and overall sustainability. Healthcare green IT efficiency improvements must be made in compliance with the expanding regulations to protect patient privacy (Wikipedia, Personal Health Record (PHR), 2017).

All of us can contribute to help improve the outlook for healthcare by contributing to the use of green IT and electric energy sustainability, in general.

Additional Reading

500-plus Hospitals join forces to Green the Healthcare Industry – http://www.greenbiz.com/blog/2012/04/05/500-plus-hospitals-join-forces-green-healthcare-industry.

Healthcare accounts for eight percent of U.S. Carbon Footprint – http://www.uchospitals.edu/news/2009/20091110-footprint.html.

Overview of Healthcare Industry, http://www.technofunc.com/index.php/domain-knowledge/healthcare-industry/item/overview-of-healthcare-industry,

Protected Healthcare Information (PHI), presentation – http://www.hhs.gov/ocr/privacy/hipaa/understanding/training/udmn.pdf.

Six Sigma, Wikipedia, http://en.wikipedia.org/wiki/Six_Sigma.

Telemedicine: barriers and opportunities in the 21st century, by B. Stanberry the Journal of Internal Medicine accessed on 8th May 2017 at http://onlinelibrary.wiley.com/doi/10.1046/j.1365-2796.2000.00699.x/full.

The Key to a Successful Telehealth Network accessed on 9th October 2016 at https://www.telushealth.co/item/key-successful-telehealth-network/.

The PHR tool – http://www.myphr.com/startaphr/what_is_a_phr.aspx.

The Role of the Healthcare Sector in Expanding Economic Opportunity, Harvard
   University publication under Economic Opportunity Series [Online]. Available:
   http://www.hks.harvard.edu/m-rcbg/CSRI/publications/report_21_EO%20
   Health%20Care%20Final.pdf.
To understand what an EHR (Electronic Health Record) contains, the URL mentioned below (accessed on 2nd November 2016) can be visited – https://www.
   healthit.gov/providers-professionals/faqs/what-information-does-
   electronic-health-record-ehr-contain.

Chapter 6
Data Storage Strategies for Green Healthcare

Success depends upon previous preparation,
and without such preparation there is sure to be failure.

—Confucius

As discussed at the end of Chap. 4, Six Sigma is a set of techniques and tools for process improvement. Their techniques, originally developed by Motorola in the 1980s, are being used by many industries throughout the world. The Six Sigma steps applicable for green IT and sustainability are:

1. Define
2. Measure
3. Analyze
4. Improve
5. Control

These five steps are similar to the five-step process that IBM has used for several years for creating energy-efficient "green" data centers: (1) diagnose, (2) manage and measure, (3) use energy-efficient cooling, (4) virtualize, and (5) build new or upgrade facilities when feasible. Details on using the Six Sigma steps were given in Chap. 4. They are repeated here since these steps are helpful in analyzing the best data-storage strategies for green healthcare.

Green Healthcare: Data Storage (SAN) Considerations

This section discusses the energy impact of using storage virtualization for servers. A typical architecture for virtual storage is storage area network (SAN) devices under a SAN Volume Controller (SVC) (Lamb, 2009).

© Springer International Publishing AG, part of Springer Nature 2018
N. S. Godbole, J. P. Lamb, *Making Healthcare Green*,
https://doi.org/10.1007/978-3-319-79069-5_6

Impacts of Server Virtualization on Data Storage

An ESG research survey of virtual server users reveals some interesting storage technology and implementation trends. The Enterprise Strategy Group (ESG) is an IT analysis and consulting firm focused on information storage, security, and management (http://www.esginc.com). This section will reference that ESG survey and relate it to some of the case studies given in later chapters. The survey stated that it is well known that virtualization is one of the hottest trends in IT. But the survey was intended to show how virtual servers impact end users' storage strategies. Virtual servers have unique requirements in the areas of performance and data protection, and users are just beginning to implement storage technologies and products that will enable them to maximize the benefits of server virtualization.

Infrastructure Options/Plans

Although one of the primary benefits of server virtualization is consolidation of resources, the implementation of virtual servers often leads to significant increases in storage capacity. Over half (54%) of the virtual server adopters have experienced a net growth in capacity, while only 7% reported a net decrease. ESG analysts believe those organizations reporting no change in capacity, or a decrease in capacity, may have benefited from storage consolidation or other infrastructure initiatives as part of their virtualization deployment and/or may simply be in the early stages of implementation and have not yet reached the tipping point where they experience a net increase in storage capacity. However, the overall conclusion is that server virtualization typically increases capacity requirements.

Together with performance and management demands, the increased storage capacity requirements have a profound effect on how users design their underlying storage infrastructures (e.g., DAS, NAS, Fiber Channel SAN, or iSCSI SAN). And in this context, it is important to note that the vast majority (72%) of users are sharing storage resources between virtualized and non-virtualized (physical) servers. Overall, there is a clear trend toward networked storage architectures (Fiber Channel SAN, iSCSI SAN, and/or NAS), as opposed to DAS. For example, 86% of the ESG survey respondents are using networked storage, while only 14% are still relying exclusively on DAS.

As might be expected, Fiber Channel SANs are preferred by larger organizations, while DAS is often the preferred architecture for SMBs. Surprisingly, however, adoption rates for iSCSI in virtual server environments are about the same across all sizes of organizations (as is the case for NAS). In any case, the trend toward networked storage in virtual server environments is clear: Today, approximately 60% of users' virtual server capacity is networked, and that percentage is expected to increase to 74% over the next 24 months. The most commonly cited benefits include better mobility of virtual machines across physical servers (66% of survey respondents), easier and more cost-effective disaster recovery, increased uptime and availability, more efficient upgrades of physical servers, and high-availability storage of multiple copies of virtual machine images (54%).

Storage Management Issues

Virtual servers force users to address storage management and data-protection issues such as backup, remote replication, capacity planning, and information security in new ways. But of all the concerns about implementing virtual server environments, performance comes out on top although, collectively, storage management issues are also of great concern.

Data Protection

It's not surprising that end users expect server virtualization and consolidation to reduce the total number of backup licenses they have to purchase. (Almost a quarter of the survey respondents report that they have been able to reduce the number of backup licenses after deploying virtual servers.) To that end, a variety of vendors are eliminating the need to have backup agents on every virtual machine. And in a related survey finding, 17% of the users have changed their backup software as a direct result of implementing server virtualization.

Disaster Recovery Planning (DRP) and Best Practices

Improving disaster recovery is one of the driving forces behind the combination of server virtualization and networked storage. In the ESG survey, 26% of the virtual server users said they are replicating virtual machines to a remote disaster recovery site, and another 39% plan to do so. One of the advantages of server virtualization is that it enables users to replicate many servers to relatively inexpensive virtual machines rather than to physical servers, which significantly reduces the primary barrier to disaster recovery: high costs. In addition, disaster recovery with virtual machines can be less costly than with physical servers because the process can in many cases be managed by the virtualization software.

A primary driving force behind remote replication in the context of server virtualization is the end users' desire to reduce their recovery time objectives (RTO). For example, 85% of the survey respondents agree that replicating virtual machine images for disaster recovery will enable their organizations to lower their RTO.

Server and Storage Virtualization

Although server virtualization and storage virtualization are usually viewed separately by IT organizations, the clear trend is toward a merging of the two technologies. The primary benefits of server virtualization include lower costs, improved
resource utilization, non-disruptive upgrades, and increased availability – all of
which are fundamentally enabled by decoupling servers, applications, and data
from specific physical assets. Storage virtualization takes those same benefits and
extends them to the underlying storage domain. Just as using networked storage for
virtual machines instead of DAS means that there is no single point of failure at a
disk system level that can bring down many virtual machines at once, storage virtualization adds yet another layer of protection against failures – extending full hardware independence from the server domain to the storage domain.

Further Details on Storage and Server Virtualization

As we've seen, computer systems are not the only candidates for virtualizing; storage can be virtualized too. This section describes the IBM SAN Volume Controller,
which provides a virtual pool of storage consisting of SAN-attached physical storage devices.

IBM SAN Volume Controller

The SAN Volume Controller (SVC) is a hardware device that brings storage devices
in a SAN together in a virtual pool. This makes your storage appear as one logical
device to manage. To the connected computers, SVC offers virtual disks as ordinary
SCSI devices. On the SAN side, SVC integrates various storage subsystems, even
multivendor, and takes care of the correct block mapping between the SAN devices
and the virtual disks for the computers. Figure 6.1 illustrates how it works.

The following points make the SVC an attractive tool for an energy-efficient
storage strategy:

- Data migration from older to newer, more efficient systems can happen
  transparently.
- Tiered storage enables you to use media with a smaller energy footprint while the
  SVC cache improves its performance.
- Consolidation of the system's individual storage devices to virtual storage has
  the same effect – increasing storage utilization – as is shown for server
  virtualization.

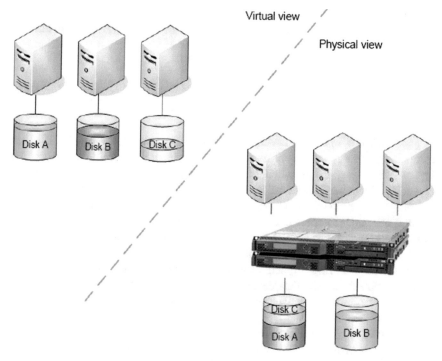

Fig. 6.1 Storage virtualization: virtual view and physical view

Storage virtualization requires more effort than server virtualization, often requiring us to rethink the existing storage landscape. During consolidation, large amounts of data must be moved from the old systems to the consolidated storage system. This can become a long task that requires detailed planning. However, when done, the effect can be enormous because now storage can be assigned to systems in the most flexible way.

Virtual Tapes

Tapes are the cheapest medium on which to store data. They offer the largest storage volume at the lowest cost, which is the reason that they are the optimal backup medium. Tapes have a long latency compared to disks. This is not always a benefit. Currently, data centers are faced with a time limitation for backing up and probably restoring their data because the time frames for backups shrink, while the amount of data to back up expands. For this reason, many sites prefer large disk-based backup systems instead of tapes.

Tape virtualization may be a solution to this problem. A virtual tape server behaves just like a tape library but a very fast one. This is made possible with internal disk arrays and a migration strategy to export to and import from real tape libraries.

Client Virtualization

A great potential in energy savings is client, or desktop, virtualization. IBM has estimated an energy savings of more than 60% by using client virtualization. In a typical workplace, the installed PCs show marginal usage rates. Except when launching an application, an office PC spends most of its time waiting for the user to press a key or click a mouse. However, the PC continues to need a considerable amount of energy to operate, heats up its surrounding environment, and produces noise. Desktop virtualization can dramatically improve the situation.

The underlying principal of client virtualization is to replace the office workstation with a box having a much smaller energy footprint. The needed computing power is moved into the data center. Today's virtualization techniques make this approach even more attractive. The benefits are many and not only to the energy balance.

Software deployment, for example, can become a mess if the desktop machine contains many historically grown software stacks. If we do not want to bother users by running updates during the day, machines can run overnight. An erroneously switched off machine can make a whole deployment fail. Central machines reduce the risk and cost.

Where to Start

First, measure energy, cooling, and inlet temperatures where possible. Trending analysis can help identify efficiency needs and opportunities. Find out what tools your systems provide. The objective is to make your energy consumption measurable. This is the most important step for all future decisions.

Aggressively drive consolidation of servers and storage using virtualization. When you are prepared for a virtualized infrastructure, consolidation comes easy. Ask the following questions:

- Which servers should be replaced in the near future? Which of the new, energy-efficient systems would be able to take on this workload because they are underutilized?
- Which systems, if any, are complementary loaded? Consider the following points:
  - Low CPU and high memory usage can nicely coexist with high CPU and low memory usage systems in a partitioned environment.

- Look for non-overlapping workloads such as day versus night jobs.
- Is any system completely filled with non-business-critical work (low performance or response time needs)? If yes, virtualize it. Pair it with another virtual system on which you run occasional peak loads, such as weekly or monthly balancing. By setting appropriate entitlements, the peak load can displace the non-critical load, while the overall system usage stays close to 100%.

• Are there individual infrastructure servers such as LDAP, DNS, and licensing? These servers are often built on a single system that then has a low utilization. If so, virtualize and consolidate to get rid of these inefficiencies. Virtualization in this case also helps in reducing the cost of redundancy. Move the virtualized image to another machine instead of providing a hot standby.

• Use virtualized storage for new and virtualized servers. Although this can require start-up costs, it pays back quickly. Instead of having terabytes of unused disk space spinning in individual machines, consolidate and rightsize your storage. Moving a virtual operating system from one machine to another is as simple as connecting the target machine with the virtualized disks.

Consider opportunities for client consolidation. The cost of managing a distributed environment of fully equipped user PCs is nearly always more than you think, so the potential for big savings exists.

A Further Look at Reasons for Creating Virtual Servers

Consider this basic scenario. You're in charge of procuring additional server capacity at your hospital's data center. You have two identical servers, each running different Windows applications for your hospital. The first server – let's call it Server A – is lightly used, reaching a peak of only 5% of its CPU capacity and using only 5% of its internal hard disk. The second server – let's call it Server B – is using all of its CPU (averaging 95% CPU utilization) and has basically run out of hard disk capacity (i.e., the hard disk is 95% full). So you have a real problem with Server B. However, if you consider Server A and Server B together, on average the combined servers are using only 50% of their CPU capacity and 50% of their hard disk capacity. If the two servers were actually virtual servers on a large physical server, the problem would be immediately solved since each server could be quickly allocated the resource each needs. In newer virtual server technologies, e.g., Unix Logical Partitions (LPARS) with micro-partitioning – each virtual server can dynamically (instantaneously) increase the number of CPUs available by utilizing the CPUs currently not in use by other virtual servers on the large physical machine. This idea is that each virtual server gets the resource required based on the virtual server's immediate need.

Figure 6.2 shows typical server utilization for stand-alone servers (i.e., NO virtualization). The multimillion dollar mainframes are typically utilized on a 24/7 basis at least partly because of the large financial investment. Mainframe "batch"

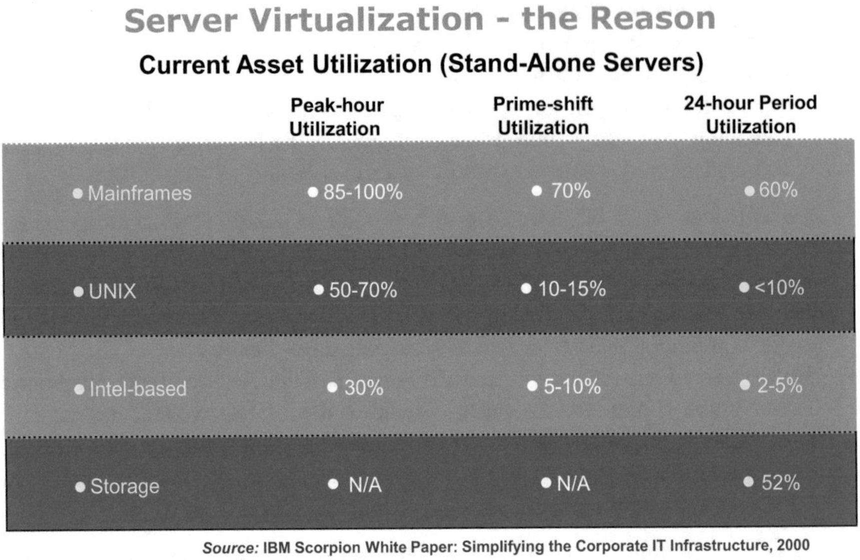

Server Virtualization - the Reason
Current Asset Utilization (Stand-Alone Servers)

	Peak-hour Utilization	Prime-shift Utilization	24-hour Period Utilization
● Mainframes	● 85-100%	● 70%	● 60%
● UNIX	● 50-70%	● 10-15%	● <10%
● Intel-based	● 30%	● 5-10%	● 2-5%
● Storage	● N/A	● N/A	● 52%

Source: IBM Scorpion White Paper: Simplifying the Corporate IT Infrastructure, 2000

Fig. 6.2   Server virtualization – the reason

processes such as running daily, weekly, and monthly corporate summary reports are typically CPU intensive and are run at night and on the weekends. The small department Windows server (labeled "Intel-based" in the diagram) is not typically used at night or on the weekends. Creating virtual servers of those Intel-based servers not only allows much better and easier sharing of resources for a mix of lightly and heavily used servers (as in the Server A/Server B example above) but also tends to spread out the utilization over 24 h on the large physical server that houses the virtual servers.

The Ultimate in Server and Data Storage Virtualization

Grid computing is a major evolutionary step that virtualizes an IT infrastructure. It's defined by the Global Grid Forum (www.gridforum.org) as distributed computing over a network of heterogeneous resources across domain boundaries and enabled by open standards. While the industry has used server cluster technology and distributed computing over networks for nearly two decades, these technologies cannot in themselves constitute grid computing. What makes grid computing different is the use of open source middleware to virtualize resources across domains.

Cloud Computing: The Future

Cloud computing is a label for the subset of grid computing that includes utility computing and other approaches to the use of shared computing resources. Cloud computing is an alternative to having local servers or personal devices handling users' applications. Essentially, it is an idea that the technological capabilities should "hover" over everything and be available whenever a user wants. Of course for healthcare uses such as mobile medicine, cloud computing must be done with adequate security. Private cloud architecture should be considered for most healthcare applications.

Cluster Architecture for Virtual Servers

As discussed earlier in this chapter, there are now many IT vendors with virtual servers and other virtual systems. Also briefly discussed was VMware's technology for moving servers called VMotion. By completely virtualizing servers, storage, and networking, an entire running virtual machine can be moved instantaneously from one server to another. I've worked with VMotion implementations on projects at data centers, and it is a step in the direction in support of the grid and cloud computing concepts discussed above.

VMware Clusters (VMotion)

The entire state of a virtual machine is encapsulated by a set of files stored on shared storage, and VMware's VMFS cluster file system allows both the source and the target VMware ESX server to access these virtual machine files concurrently. The active memory and precise execution state of a virtual machine can then be rapidly transmitted over a high-speed network. Since the network is also virtualized by VMware ESX, the virtual machine retains its network identity and connections, ensuring a seamless migration process. VMotion servers require external shared storage (SAN).

Figures 6.3 and 6.4 show the VMotion concept and four node cluster example implemented at a large data center outside of Chicago. The concept of using VMotion clusters for all VMware server needs has worked very well and has basically eliminated the previous need to manage and balance the load among different physical servers used to host VMware virtual servers. Figure 6.3 shows the general VMware cluster (VMotion) concept used at this data center. Figure 6.4 shows one of the two clusters used.

The following list gives the details of the VMware/VMotion architecture used for the data center.

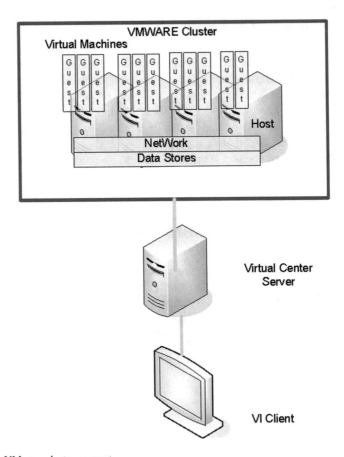

Fig. 6.3 VMware cluster concept

- VMware ESX operating systems.
- VMware Virtual SMP – VM to use up to four physical processors simultaneously.
- VMware High Availability (HA) for any application running in a virtual machine, regardless of its operating system or underlying hardware configuration.
- VMware Distributed Resource Scheduler (DRS) monitors utilization across resource pools and allocates available resources among the virtual machines based on predefined rules.
- VMware VMotion. The entire state of a virtual machine is encapsulated by a set of files stored on shared storage, and VMware's VMFS cluster file system allows both the source and the target VMware ESX server to access these virtual machine files concurrently.
- VMware VMFS – VMFS is a cluster file system that leverages shared storage to allow multiple instances of VMware ESX to read and write to the same storage.
- VMware Consolidated Backup – LAN-free backup.

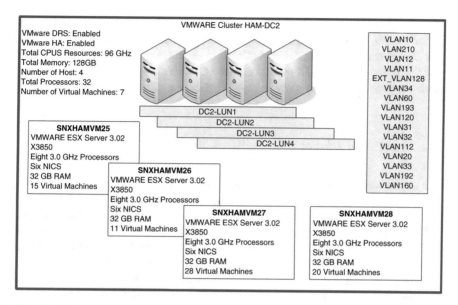

Fig. 6.4 VMware cluster example with four nodes

- VMware Update Manager – automation of patches and updates.
- VMware Storage VMotion – enable live migration of virtual machine disk files across storage arrays with VMware Storage VMotion.

Blade Servers and Virtualization

Two of the more recent trends in data-center optimization are the adoption of blade servers and the deployment of server virtualization. Each of these can bring benefits to healthcare. What's more, combining blade servers with a virtualization strategy can lead to even greater efficiencies for enterprise data centers. This section will compare and contrast the two technologies that are both used for data-center simplification and energy savings.

A blade server is a chassis housing that contains multiple, modular electronic circuit boards (blades), each of which includes processors, memory, storage, and network connections and can act as a server on its own. The thin blades can be added or removed, depending on needs for capacity, power, cooling limits, or networking traffic. The products, which are designed for high-density computing, can provide a host of benefits to organizations – including more efficient use of space and energy.

Blade servers provide a broad, modular platform that works well for consolidation as well as building for the future. Blades provide an opportunity for a reduction in complexity by reducing the number of IT components, due to shared components in the architecture. There should be greater manageability, modularity, and flexibility for growth. These all help in reducing TCO (total cost of ownership). With blades you also provide a good modular platform for the future architecture that will support things like I/O, virtualization, and ease of provisioning. Blades should be considered a good hardware counterpart to virtualization software.

The Benefits of Blades

One of the key IT goals of many organizations today is to economize on space, power, and cooling in the data center. In general, blade servers provide a smaller form factor and a better footprint than other types of servers, so organizations can make better use of their data centers. There has definitely been a greater need for this type of technology as energy costs continue to go up. Other benefits of blade servers are more specific to individual manufacturers. For example, some suppliers provide strong management tools. The blade server devices include cutting-edge technology that allows hospitals and their IT staffs to address fundamental data-center challenges, such as cost, change, time, and energy, in order to achieve better business outcomes. Blade servers are flexible in terms of expansion and management, which is a great benefit for administrators. Blades can also be used in a complete blade system to bring together technologies like virtualization, automation, energy efficiency, and unified management.

Blade servers are available from many vendors, including Hewlett-Packard, Dell, and IBM. Sales of blade server technology continued to accelerate in the 2017 time period according to research firm IDC. Overall, blade servers, including x86, EPIC (explicitly parallel instruction computing) and RISC (reduced instruction set computer) blades, currently account for about 20% of server market revenue (2016 data). However, blade server factory revenue continues to grow rapidly, according to IDC.

Virtualizing Blade Servers?

As we've discussed previously in this chapter, server virtualization enables servers to be converted into pools of logical computing resources. Through the use of virtualization software, a single server can be logically divided up into multiple virtual devices. This enables the same hardware device to run multiple operating systems and applications – each independently of the others – as if they were on physically separate devices.

As we've also discussed in this chapter, server virtualization offers a host of potential benefits. Virtualization more efficiently employs IT resources to improve asset utilization and simplify management of the data center. It can help improve operational availability by providing flexible resources and software tools that automatically assess when servers need additional resources and adjust capacity in real time. This alleviates potential bottlenecks and the slowing of systems due to over provisioning.

Hospital Continuity-Disaster Recovery

Beyond cost savings, disaster recovery, business continuity, and data recovery are other reasons hospitals and other healthcare organizations are using server virtualization. The same technology is used for virtualizing desktops that go into the hospital's data center. Many hospitals have two physical PCs for one person. Virtualization can eliminate one of those. You can run multiple environments on a single device. Developers, in particular, need multiple environments. Virtualization allows you to take a desktop PC off a person's desk and move it into a data center. There are a lot of manageability benefits to this. The key use cases that are most commonly mentioned for server virtualization are server consolidation, business continuity and disaster recovery, software lab automation, and the desktop scenarios. The two ways to leverage virtualization for desktops include running virtualization on the desktop so that you can have multiple environments on a single system and running desktop environments on servers accessed via thin clients – what we refer to as virtual desktop infrastructure or VDI. VDI helps organizations by making it easier for them to manage, secure, and protect desktop environments. Developers and call centers are two examples of where hospitals have been deploying VDI solutions. Another benefit of server virtualization is increased agility. Virtualization solutions help make IT environments more flexible by simplifying the management and automation of virtual infrastructures.

HP has been very vocal on this point. HP indicates that businesses can speed up the deployment of infrastructure components and applications, keep them up and running more efficiently, and adjust infrastructure more quickly when business demands change. Improved quality of service is yet another business benefit. Virtualization products can improve the quality of IT service delivery by aligning IT supply with business demand. Server virtualization can help mitigate business risk if leveraged for business continuity and disaster recovery purposes. When combined with the right set of management tools, which varies depending on the environment, virtualization can provide a very affordable alternative to maintaining separate duplicate sites for disaster recovery.

Combining Blades and Virtualization

Deploying both blade servers and virtualization can lead to even greater efficiencies in the data center. Physical data-center space, power and cooling needs, and costs are all reduced when a virtualization strategy is supported by the use of blade servers. Other benefits include increased server utilization rates and increased reliability, flexibility, and serviceability. Blades and virtualization work well together to provide greater manageability and modularity on both the hardware side and the software side. Both address space-saving issues and consolidation of underutilized servers and ease of provisioning new servers. They address many of the same issues and are a good hardware and software combination.

Blades and virtualization can be very highly complementary strategies. Many users are deploying both at the same time. They work together to provide higher-density computing, giving you a lot more out of your data-center blades due to the much smaller form factor and the ability to consolidate multiple workloads on fewer servers. Both also work together to reduce energy consumption costs and conserve data-center space. They also make it easier to manage the data center. The combination of blades and virtualization can lead to clear business benefits. These are both technologies that go directly to ROI (return on investment) and reducing TCO. You get more for what you're spending. Lower TCO comes through reductions in both capital expenses and operating expenses.

Another plus is that server blades and virtualization can help healthcare organizations make their data centers more environmentally friendly. For example, there are many cases where hospitals would have needed to build a new data center to support growth, had they not implemented a virtualization strategy. By using virtualization, they've avoided the cost of having to build a big new data center. A typical example is a hospital that has perhaps 300 servers that it can't bring up because the hospital doesn't have enough power in its data center. That hospital needs to economize. They can eliminate the need for most of those 300 additional servers with virtualization and blades. Virtualization with blade servers allows IT administrators to spend less time managing and reorganizing the data center. The time savings alone gives staff the opportunity to explore new ways to help their hospitals grow from an IT perspective. Based on experience, the time it takes an administrator to bring up a new server in a data center is substantial. In a virtualized environment, the administrator can bring up a new server from his or her desk in a matter of minutes. IT staff can invest in more productive uses of their time.

Both blades and virtualization are very strong enabling technologies for the level of advanced management that will move us closer to dynamic IT or utility computing. They offer the ability to pick the computing resources, the network resources, and the storage resources needed for each project.

Impacts of Server Virtualization on Data Storage

An ESG research survey of virtual server users reveals some interesting storage technology and implementation trends. The Enterprise Strategy Group (ESG) is an IT analyst and consulting firm focused on information storage, security, and management (http://www.esginc.com). This section will reference that ESG survey and relate it to some of the case studies given in later chapters. The survey stated that it is well known that virtualization is one of the hottest trends in IT. But the survey was intended to show how virtual servers impact end users' storage strategies. Virtual servers have unique requirements in the areas of performance and data protection, and users are just beginning to implement storage technologies and products that will enable them to maximize the benefits of server virtualization.

Infrastructure Options/Plans

Although one of the primary benefits of server virtualization is consolidation of resources, the implementation of virtual servers often leads to significant increases in storage capacity. Over half (54%) of the virtual server adopters have experienced a net growth in capacity, while only 7% reported a net decrease. ESG analysts believe those organizations reporting no change in capacity, or a decrease in capacity, may have benefited from storage consolidation or other infrastructure initiatives as part of their virtualization deployment and/or may simply be in the early stages of implementation and have not yet reached the tipping point where they experience a net increase in storage capacity. However, the overall conclusion is that server virtualization typically increases storage capacity.

Together with performance and management demands, the increased storage capacity requirements have a profound effect on how users design their underlying storage infrastructures (e.g., DAS, NAS, Fiber Channel SAN, or iSCSI SAN). And in this context, it's important to note that the vast majority (72%) of users are sharing storage resources between virtualized and non-virtualized (physical) servers. Overall, there is a clear trend toward networked storage architectures (Fiber Channel SAN, iSCSI SAN, and/or NAS), as opposed to DAS. For example, 86% of the ESG survey respondents are using networked storage, while only 14% are still relying exclusively on DAS.

As might be expected, Fiber Channel SANs are preferred by larger organizations, while DAS is often the preferred architecture for SMBs. Surprisingly, however, adoption rates for iSCSI in virtual server environments are about the same across all sizes of organizations (as is the case for NAS). In any case, the trend toward networked storage in virtual server environments is clear: Today, approximately 60% of users' virtual server capacity is networked, and that percentage is expected to increase to 74% over the next 24 months. The most commonly cited benefits include better mobility of virtual machines across physical servers (66% of survey respondents), easier and more cost-effective disaster recovery, increased uptime and availability, more efficient upgrades of physical servers, and high-availability storage of multiple copies of virtual machine images (54%).

Storage Management Issues

Virtual servers force users to address storage management and data-protection issues such as backup, remote replication, capacity planning, and information security in new ways. But of all the concerns about implementing virtual server environments, performance comes out on top although, collectively, storage management issues are also of great concern.

Data Protection

It's not surprising that end users expect server virtualization and consolidation to reduce the total number of backup licenses they have to purchase. (Almost a quarter of the survey respondents report that they have been able to reduce the number of backup licenses after deploying virtual servers.) To that end, a variety of vendors are eliminating the need to have backup agents on every virtual machine. And in a related survey finding, 17% of the users have changed their backup software as a direct result of implementing server virtualization.

Disaster Recovery

Improving disaster recovery is one of the driving forces behind the combination of server virtualization and networked storage. In the ESG survey, 26% of the virtual server users said they are replicating virtual machines to a remote disaster recovery site, and another 39% plan to do so. One of the advantages of server virtualization is that it enables users to replicate many servers to relatively inexpensive virtual machines rather than to physical servers, which significantly reduces the primary barrier to disaster recovery: high costs. In addition, disaster recovery with virtual machines can be less costly than with physical servers because the process can in many cases be managed by the virtualization software.

A primary driving force behind remote replication in the context of server virtualization is the end users' desire to reduce their recovery time objectives (RTO). For example, 85% of the survey respondents agree that replicating virtual machine images for disaster recovery will enable their organizations to lower their RTO.

Server and Storage Virtualization

Although server virtualization and storage virtualization are usually viewed separately by IT organizations, the clear trend is toward a merging of the two technologies. The primary benefits of server virtualization include lower costs, improved resource utilization, non-disruptive upgrades, and increased availability – all of which are fundamentally enabled by decoupling servers, applications, and data from specific physical assets. Storage virtualization takes those same benefits and extends them to the underlying storage domain. Just as using networked storage for virtual machines instead of DAS means that there is no single point of failure at a disk system level that can bring down many virtual machines at once, storage virtualization adds yet another layer of protection against failures – extending full hardware independence from the server domain to the storage domain.

Chapter Summary and Conclusions

The following conclusions can be made from the above discussion on virtualization and green IT:

- Virtualization is the most promising technology to address both the issues of IT resource utilization and facilities space, power, and cooling utilization.
- Many IT companies are addressing the situation from end to end, at the server end through power management features and at the data-center ends through integrated IT/facilities modular solutions.
- IT virtualization includes server virtualization, storage virtualization, client virtualization, virtualization using cluster architecture, and virtualization of blade servers.
- The ultimate objective and benefit of virtualization is the significant IT flexibility it brings to corporate users. IT virtualization can benefit data protection, business continuity, and disaster recovery.

Additional Reading

Koomey, Jonathan G. Ph.D. "*Data center electricity use*" Lawrence Berkeley National Laboratory & Stanford University. Presented at the EPA stakeholder workshop at the Santa Clara Convention Center, February 16, 2007.

Mitchell, Robert. "*Seven steps to a green data center*", Computerworld, April 21, 2007.

Molloy, Chris. "*Project big green: The rest of the story*", IBM Technical Leadership Conference (TLE), Orlando, Florida, April, 2008.

Obama, Barack. "*The irreversible momentum of clean energy*", Science Magazine, January 13, 2017.

Pacific Gas & Electric (PG&E), "*Data center energy management*", http://hightech.lbl.gov/DCTraining/, 2008.

Patterson, M.K., Costello, D., Grimm P, Loeffler, M. "*Data center TCO; A comparison of high-density and low-density spaces*," THERMES 2007, Santa Fe, NM, 2007.

Pope Francis. "*Laudato Si – on care for our common home*", July 18, 2015

Rassmussen, N. *"Electrical efficiency modeling of data centers,"* White Paper #113, APC. http://www.apcmedia.com/salestools/NRAN-66CK3D_R1_EN.pdf, 2005.

Rodriguez, Jean-Michel. *"Green data center of the future – architecture"*, IBM Technical Leadership Conference (TLE), Orlando, Florida, April, 2008.

The Green Grid. *"The green grid data center power efficiency metrics: PUE and DCiE"*, The Green Grid, http://www.thegreengrid.org/gg_content/TGG_Data_Center_Power_Efficiency_Metrics_PUE_and_DCiE.pdf , 2007.

Velte, T. J., Velte, A. T., Elsenpeter, Robert. *"Green IT: Reduce your information system's environmental impact while adding to the bottom line"*, McGraw-Hill, ISBN: 978-0-07-159923-8, 2008.

Chapter 7
Collaborate to Maximize Reduction of Hospital and Healthcare Energy Use and Carbon Footprint

According to a 2012 study from the University of Illinois-Chicago's School of Public Health, the hospital industry could save $5.4 billion in 5 years and up to $15 billion in 10 years if it adopts sustainable practices.

In order to meet the challenge for effective green healthcare, collaboration is a must. Data-center technology for green healthcare covers a broad spectrum from efficient cooling towers and variable speed blowers to the use of energy-efficient IT systems such as virtual servers, blade centers, and virtual data storage. Significant contributors to the collaboration team include IT technology vendors, data-center design businesses, infrastructure technology providers, energy utilities, and governments. This chapter includes descriptions on how to help make this collaboration happen. A key starting point is to have an executive champion for implementing a green healthcare center. Energy utilities have additional interest in implementing green IT since they can use their experience to help establish rate-case incentives on green computing technology for their customers.

The previous chapters have mentioned some of the many groups that are involved in green IT. The following sections give additional information on how many of these groups are collaborating with healthcare providers.

IT Technology Vendors

There is evidence that the continued focus on energy-efficient computing has forced tech giants in Silicon Valley and other areas to collaborate more than ever before. This collaboration should result in a better range of energy-efficient products (hardware and software) as well as faster product turnaround. Companies are collaborating by discussing what works and what doesn't work instead of everybody having to reinvent the wheel individually.

© Springer International Publishing AG, part of Springer Nature 2018
N. S. Godbole, J. P. Lamb, *Making Healthcare Green*,
https://doi.org/10.1007/978-3-319-79069-5_7

For example, competitors like HP and IBM are sitting at the same table to work out green standards with The Green Grid. Department of Energy (DOE) laboratories such as Lawrence Berkeley National Labs and Pacific Northwest National Labs are cooperating in development of research and best practices for green IT. Government agencies like the Environmental Protection Agency (EPA) are leading the effort to standardize metrics to measure energy consumption. The EPA's 2007 report to congress on data-center energy efficiency (EPA, 2007) helped push the Energy Star program on energy efficiency for all electrical devices. Organizations are taking on the responsibilities that they can deliver on – not worrying about turf.

Data-Center Design-to-Build Businesses

There are several industry groups focused on data-center energy efficiency. The Green Grid is a significant group in this area (The Green Grid, n.d.). Its primary mission is to further data-center energy efficiency initiatives. The Green Grid is made up of IT equipment and component (power supplies, microprocessors, etc.) manufacturers, manufacturers of facility support equipment (CRACs, UPS, rack-level cooling devices, and others), and data-center operators and designers. The group is seeking to drive the development of metrics which provide a meaningful measure of data-center energy utilization and performance; drive the development of standard protocols to facilitate collection, sharing, and reporting of power and thermal data from IT and facilities equipment; and design operating strategies which promote optimal data-center energy use. The Green Grid has published a collection of white papers on data-center metrics and data management. Please see http://www.thegreengrid.org/ for additional information on The Green Grid.

Also, ASHRAE (American Society of Heating, Refrigerating, and Air-Conditioning Engineers) has had a data-center subcommittee in operation for many years. The group publishes a set of documents and manuals which provide data centers with cooling strategies which optimize cooling delivery, data-center temperature, and humidity profiles and maximize the cooling delivered per unit of energy applied. Please see http://www.ashrae.org/ for additional information on ASHRAE.

The Uptime Institute consortium of data-center operators focuses on tools and strategies to maintain the serviceability and reliability of data centers. The Institute sets various data-center performance levels through the tier system, which dictates equipment redundancy and data-center environment criteria to maintain a specified level of system availability. Uptime has been adding energy optimization as well as other critical criteria for data-center operation and reliability and one that needs to be factored in when considering data-center reliability and availability. Please see http://uptimeinstitute.org/ for additional information on the Uptime Institute.

Another group that is involved in data-center and energy efficiency activities is AFCOM (Association for Computer Operations Management), which provides education and resources for data-center managers. They have been devoting a portion of their efforts to assisting data-center managers to identify energy efficiency activities. Please see http://www.afcom.com/ for additional information on AFCOM.

Many other groups are associated with energy efficiency activities. One very active group is the Alliance to Save Energy (ASE). The ASE mission: "The Alliance to Save Energy promotes energy efficiency worldwide to achieve a healthier economy, a cleaner environment, and greater energy security. Energy efficiency is the quickest, cheapest, cleanest way to extend our world's energy supplies." See http://www.ase.org/ for additional information on the ASE. In addition to the groups discussed previously in this section, the ASE collaborates with the following organizations:

- *ITherm:* "An International conference for scientific and engineering exploration of thermal, thermo-mechanical, and emerging technology issues associated with electronic devices, packages, and systems." (www.itherm.com)
- *Efficient Power Supplies:* A web site created by EPRI Solutions Inc. and Ecos Consulting to encourage a global discussion of energy-efficient power supplies. (www.efficientpowersupplies.org)
- *Consortium for Energy Efficient Thermal Management:* A Georgia Institute of Technology consortium to conduct "research on thermal and energy management of electronics and telecommunications infrastructure." (http://www.ceetherm.gatech.edu/)
- *7x24:* An association facilitating the exchange of information for "those who design, build, use, and maintain mission-critical enterprise information infrastructures…7x24 Exchange's goal is to improve the end-to-end reliability by promoting dialogue among these groups." (http://www.7x24exchange.org/index.html)

Collaboration of Building Energy Management and IT Energy Management

Beyond the familiar challenge of establishing energy-efficient data centers for healthcare lies a huge opportunity scarcely tapped by IT: the green possibilities of the hospital building itself. Growth is driving global trends in resource depletion, air and water pollution, energy consumption, and climate change. A third of US energy consumption comes from commercial buildings. Businesses are automating those buildings to reduce costs and emissions. IT can have a green impact on a hospital's energy and emissions: start with the data center; manage desktop energy use; and enable mobility.

IT departments operate in an environment surrounded by sophisticated data acquisition, analysis, and networking systems of which IT itself is largely unaware. Building automation systems (BAS) are the brains of commercial and industrial buildings that control their own environments. The benefits of building automation are energy savings, improved occupant comfort, added security and safety, and reduced maintenance costs. All of these benefits are at the top of the list for conservation-minded building owners.

Building automation systems, such as lighting and temperature controls, are common in larger facilities. Energy management systems (EMS) go further, centralizing the control of lighting, heating, ventilating, and air conditioning with the goal of reducing the energy those systems consume. Almost every campus (corporate, medical, or academic) has an EMS, as do most of the Fortune 100. Manufacturers have adopted automation for efficiency, and those industrial systems are now being leveraged to reduce energy consumption.

Groups of forward-looking vendors have begun to think about how the EMS and IT worlds should converge. The concepts center around removing the long-standing wall between building networks and IT (tenant) networks. Mixed into this dialog are other low-profile systems common in most buildings, such as security, air quality, and life safety. Cisco has approached the building control industry with the notion that information is the "fourth utility" after electricity, gas, and water. Cisco has proposed moving EMS to the IP network, not only for efficiency but also for the information synergies involved. Business information has a strategic and tactical value, and information about the building's performance is no different.

Protocols, however, are among the stumbling blocks. Building systems operate on largely special-purpose open systems (such as BACnet or LonWorks), and a few proprietary systems remain popular. Today, both types of systems can talk to the IP network through gateways. Within the last few years, the building control industry has discovered XML. Middleware applications gather information and normalize it for consumption by ERP, accounting, and other enterprise applications.

Energy Utilities

Electric utilities provide very interesting case studies since they can provide incentives for their customers to move to green IT. For example, the Pacific Gas and Electric Co. (PG&E) was the first company to offer incentives for power-saving technologies, encouraging customers to get rid of underutilized computing and data storage equipment through virtualization. In addition, the company spearheaded a coalition of utilities to discuss and coordinate energy efficiency programs for the high-tech sector, focusing on data centers.

The online business tools offered by Pacific Gas and Electric Company provide companies with help to make their data centers more efficient. The tools include:

- Business tools features
- Energy usage
- Billing history
- Rate comparison tools
- Energy outage and restoration status
- Billing details
- Account aggregation

Check with your electric utility on energy audits. Some utilities such as PG&E offer free energy audits.

Of course, after your free audit, you may be eligible for rebates for your green IT initiatives. For example, see http://www.pge.com/mybusiness/energysavingsrebates/rebatesincentives/.

Governments

With all the recent publicity on the growth of energy use by enterprise-level IT equipment, let's look back to when much of the measurements began. In 2008, a report to the US Congress (Koomey, 2007 February 16) indicated that the 2008 data-center energy use – about 2% of 2008 global energy use – was expected to have double digit increases in energy growth for the next 5 years. Based on that information, various governments around the world took actions to encourage data-center operators to improve their energy performance of their data centers. Great strides in data-center energy efficiency were made in those 10 years. If we look at all current (2018) IT energy use (including the energy used for all of our laptops), then the portion of IT global energy use has now reached 10% of all global electricity use (Clark, 2013).

Government initiatives include:

- *US EPA Energy Star Data-Center Rating System*

As mentioned earlier, US EPA has an Energy Star building program. They also have a data-center rating system. The following sections give information on US DOE and EPA initiatives.

- *DOE Save Energy Now – Data-Center Initiative*

Also, as mentioned earlier, the Department of Energy is partnered with the Lawrence Berkeley National Lab to develop a model that characterizes the power use and thermal profile of the data center. The software tool will collect specific energy use data for the data center, calculate the DCIE metric, create estimated energy use breakouts by system, and prepare a list of applicable energy-saving actions. DOE had both of these efforts up and running at the end of 2008.

- *EU Code of Conduct for Data Centers*

The European Code of Conduct for Data Centers has moved forward in earnest. Here's a link to the 2017 EU code: (EU Code of Conduct, 2017).

- *Other Geographies*

Australia has been very busy in their data-center energy efficiency efforts. Overall, industry information on system power demands, utilization, and opportunities for energy efficiency improvements in data centers have made it clear that there are significant worldwide opportunities to reduce energy usage in data centers. In turn, they are promoting energy efficiency programs to encourage public and private entities to capture those opportunities.

Collaboration Within Your Own Company

As discussed above, green IT collaboration includes governments, IT vendors, electric utilities, and many other groups. However, collaboration is also needed among the different departments in your healthcare facility. The Uptime Institute recommends that every data center look at the following five issues for both short- and long-term energy savings. These are:

- Server consolidation, configuration, virtualization
- Enabling "power-save" features on servers
- Turning off "dead" servers (no longer in use but running)
- Pruning "bloatware" (the application efficiency issue)
- Improving the Site Infrastructure Energy Efficiency Ratio

Typically, data centers can improve energy savings by 25–50% over a 2-year period just by tackling each of these challenges in a cross-discipline way. The Institute has developed a multifunctional team methodology known as ICE (Integrated Critical Environment) to provide both the business and technical rigor required. A tried and true method not on this list above is to send an email to all users proposing server shutdown for maintenance for a 24 h period. Active server users complain instantly with reasons why their servers can't be shut down. If no angry responses result, shut them off indefinitely (35 days? At least for more than a single accounting period...), and then disconnect the servers. This frees up space, energy, and manpower.

In reexamining, retrofitting, and redesigning data centers, mitigating business risks are as important as energy savings. Such considerations are at the nerve center of every healthcare facility, large or small. Active participation, support, and collaboration are required from five key individuals across the organization: representatives from the offices of the CFO, CIO, real estate and facilities, data-center IT and facilities managers, and the technical teams who deal with applications and IT solutions for your company.

Universities Collaborate

Universities are in a unique position to collaborate on green IT. Columbia University in New York City is an excellent example of a university collaborating on green IT within the university, with other universities, with New York State organizations, and with New York City. Columbia's Business School's Green Club has shown its enthusiasm to collaborate in the green IT exercise. Columbia submitted its green data-center results to Educause, NYSERNet, NYSgrid, the Center for IT Leadership, the Ivy Plus consortium, and the Common Solutions Group as a real-world case study. The opportunity to rigorously measure recommended best practices and technological innovations in a real-world environment, validated by the scrutiny incorporated from the beginning via the three user groups, can have a far-reaching impact within and beyond Columbia. The Columbia Green IT collaboration also included New York City and, back in 2008, Mayor Michael Bloomberg's 10-year plan for New York City on reducing carbon emissions by 30% based on 2007 levels. Columbia University committed to that 30% reduction even in the face of greatly increased growth in high-performance computing (HPC) requirements fueled by the worldwide research community.

In the past several years, high-performance computing (HPC) has been growing at every research university, government research laboratory, and high-tech industry in New York State, nationally and internationally. HPC is a cornerstone of scientific research disciplines, many of which had previously used little or no computing resources. Researchers are now performing simulations, analyzing experimental data from sources such as the Large Hadron Collider at CERN, genetic sequencers, scanning-tunneling electron microscopes, econometric and population data, and so on. These applications have led to an explosion of computing clusters now being deployed throughout Columbia, as well as in peer research institutions, and New York State's biotechnology, nanotechnology, financial, and other industries; this increase frequently requires construction of new server rooms, putting pressure on space in existing data centers and leading to increased demand for energy. Without this research, New York State, or any other government anywhere, cannot compete in an increasingly high-tech, computationally intensive world.

The Green Grid Collaboration Agreements

The Green Grid (mentioned at the beginning of this chapter) represents an excellent example of organizations collaborating for energy efficiency. The Green Grid is a global consortium dedicated to advancing energy efficiency in data centers and business computing ecosystems. Back in 2008, The Green Grid announced memorandums of understanding (MOUs) with the US Environmental Protection Agency (EPA) and the Storage Networking Industry Association (SNIA). The Green Grid's agreement with the EPA was to first promote energy efficiency in EPA computer facilities and then broadly share results in order to impact change within both other governmental agencies and the private sector. The alliance with SNIA, formally announced by SNIA in mid-2008, was designed to further networked storage best practices for energy efficiency.

The Green Grid's memorandums of understanding with the EPA and SNIA highlight the organization's continuing efforts and progress in working with government agencies and key industry players to define and promote the adoption of standards, processes, measurements, and technologies for energy efficiency in the data center.

The Green Grid's collaboration with the EPA will accelerate the adoption of best practices for energy efficiency in existing computer/server rooms throughout the EPA. As an initial step, The Green Grid and the EPA will identify an existing small EPA computer/server room as a target for an energy efficiency showcase and execute a public project demonstrating the feasibility, approach, and benefits of optimization. The results, best practices, and real-world takeaways from this project will be shared with other governmental agencies, industry stakeholders, and the private sector. As announced by The Green Grid, this "agreement between the EPA and The Green Grid will build further collaboration between the private and public sectors, and to set an example by improving energy efficiency in federal government computer facilities. The Green Grid will play a key role in this project by providing a team of technical experts who will perform the assessment and direct strategies and techniques for maximizing energy efficiency."

SNIA also announced a formal alliance with The Green Grid. The Green Grid will work with SNIA and its Green Storage Initiative to develop and promote standards, measurement methods, processes, and technologies to improve data-center energy efficiencies. SNIA will use its expertise in networked storage and membership (over 400 member companies and 7000 individual members) to work with The Green Grid on best storage practices for achieving more efficient storage infrastructures, including more efficient storage networking technologies.

The SNIA strongly believes that addressing the challenges associated with energy efficiency and green computing will require collaboration across all IT areas, including the storage industry. The SNIA believes that The Green Grid is a key industry organization for improving energy efficiency within data centers and business computing ecosystems. Through its Green Storage Initiative, which is dedicated to applying the technical and educational expertise of the storage industry to develop and find more energy-efficient solutions in the data center, SNIA is committed to work with The Green Grid to develop best practices and education for the industry.

Collaboration and Carbon Trading

Carbon trading is an interesting area of collaboration between companies in the green space. Governments usually have the role of regulating carbon emissions through systems such as "cap and trade." In a cap and trade system, the regulatory body sets a limit or cap on the amount of pollutant a company can emit. Companies are given a limit or cap on emissions, and if they need to exceed that limit, they can buy credits from others who are emitting less than their cap. A purchase of credits is the trade.

IT Vendors and Collaboration

Back in 2008, during a big push to help establish data-center energy efficiency, IBM announced new energy management software, an expansion of its energy certificates program, and an energy benchmark to help clients establish energy efficiency goals and optimize energy efficiency, energy measurement, and verification of green IT progress across the enterprise. The offerings for energy measurement included IBM Active Energy Manager software to measure power usage of key elements of the data center, from IT systems to chilling and air-conditioning units; an expansion of IBM's energy certificates program to 34 countries; and an online energy assessment benchmark.

Energy Manager Software

IBM Systems Director Active Energy Manager (AEM) tracks energy consumption in data centers and helps customers monitor power usage and make adjustments to improve efficiency and reduce costs. The new software allows IT managers to control – even set caps on – their energy use for servers, storage, and networking as well as the air-conditioning and power management systems that keep the data center running. The software supports monitoring of devices that are connected to selected "smart" power strips used to provide power to multiple devices.

Additionally, the software can be used with equipment from facility management providers. For example, the software can retrieve temperature and power information using SynapSense Corporation's wireless sensors, which can be located virtually anywhere in the data center. It can also receive alerts and events related to power and cooling equipment through interaction with Liebert SiteScan from Emerson Network Power. The alerts can notify IT administrators about issues with facilities equipment, such as overheating, low battery power on uninterruptible power supply batteries, or other conditions that may keep IT equipment in a data center from running properly.

Global Significance of Energy Efficiency Certificates Program

To help clients benchmark and improve the efficiency of their IT operations and reduce their environmental impact, IBM and Neuwing Energy have expanded the Energy Efficiency Certificate (EEC) program to reach customers in 34 countries. This program enables clients to measure their energy usage while earning energy efficiency certificates for reducing the energy used to run their data centers. The certificates earned – based on energy use reduction verified by a certified third party – provide a way for businesses to attain a certified measurement of their energy use reduction, a key emerging business metric. The certificates can be traded for cash on the growing energy efficiency certificate market or otherwise retained to demonstrate reductions in energy use and associated CO_2 emissions.

In addition to the United States, Canada, and Mexico, clients in the following countries can now apply for energy efficiency certificates associated with improvement in IT: Ireland, the United Kingdom, France, Germany, Italy, Spain, Belgium, the Netherlands, Denmark, Portugal, Luxembourg, UAE, Saudi Arabia, Kuwait, Bahrain, Oman, Qatar, Egypt, Jordan, Pakistan, India, China, Singapore, Malaysia, Indonesia, South Korea, Thailand, Australia, New Zealand, the Philippines, and Japan.

"Establishing a worldwide energy certificates program with the help of IBM is important to clients around the globe who are dramatically improving the efficiency of their infrastructures to meet their environmental responsibility goals as opposed to simply buying renewable energy certificates," said Matthew Rosenblum, CEO and president, Neuwing Energy. "This program gives clients the incentive to become more efficient at the source and helps reduce energy costs at the same time. We have already seen dramatic results from both utilities and Fortune 500 companies as they start to understand how productive this program is in keeping economic expansion growing while reducing energy costs."

Al Gore and Green Collaboration

Green collaboration covers a wide spectrum. Nobel Prize winner and former Vice President Al Gore emphasized the need for everyone, worldwide, to help solve the climate crisis with his book and presentations on "An Inconvenient Truth" in 2007 (Gore, 2007). Ten years later in 2017, Al Gore published "An Inconvenient Sequel," an upbeat book describing all of the accomplishments in the past 10 years to help combat climate change. In April 2008, the California firm of Kleiner, Perkins, Caufield & Byers (KPCB) and Generation Investment Management announced a collaboration to find, fund, and accelerate green business, technology, and policy. At that time the firm also announced that Al Gore had joined KPCB as a partner. Back in 2008, Gore stated that the alliance would bring together world-class business talent to focus on solving the climate crisis. He emphasized that, together, KPCB and Generation have a working understanding of this urgent, multidimensional challenge and are resolved to help business and government leaders accelerate the development of sustainable solutions. The collaborating groups said the alliance represents "a landmark alignment of resources to effect global change to protect the environment. It combines the research expertise of both organizations with a track record of successful investments in public and private companies, from early stage to large capitalization business. It aligns the convening power of Mr. Gore, the KPCB Greentech Innovation Network and the Generation Advisory Board towards a common goal. In addition, KPCB's presence in Asia and the U.S., combined with Generation's presence in the U.S., Europe and Australia, will support global scale solutions."

Gore also announced that as part of the agreement between the two firms, 100% of his salary as a partner at KPCB will be donated directly to the Alliance for Climate Protection – the nonpartisan foundation he chairs that focuses on accelerating policy solutions to the climate crisis. So collaboration in the green space, including green IT, will continue to be far-reaching.

Al Gore's Plans to Save the Planet

As noted above, Al Gore has been working on combating climate change for decades. His 2017 book, "An Inconvenient Sequel: Truth to Power" (Gore, 2017), is very optimistic on the progress made in fighting climate change. In July 2008, Al Gore announced an initiative to save the planet. The plan was so bold that the July 18, 2008, issue of *Time Magazine* ran an article titled "Gore's Bold, Unrealistic Plan to Save the Planet" (Walsh, 2008). Gore challenged America to generate 100% of our electricity from sources that do not lead to global warming – and to do it within 10 years. Speaking in Washington on July 17, 2008, Gore called on Americans to completely abandon electricity generated by fossil fuels within 10 years and replace them with carbon-free renewables like solar, wind, and geothermal. It is a bold plan, almost to the point of folly. But at the very least, it's one that certainly matches the scale of his rhetoric. "The survival of the United States of America as we know it is at risk," he said. "The future of human civilization is at stake." Gore ended his speech on his plan with a rousing reminder of President John F. Kennedy's challenge to put a man on the moon – a challenge that was met, Gore noted, in less than a decade. He stated: "We must now lift our nation to reach another goal that will change history."

Gore's bold 2008 plan fits very well with the topic of this chapter: we must all collaborate on green healthcare, and much more collaboration is required to solve the issues of climate change and global greenhouse gas emissions. Gore's emphasis on problems of national security, foreign oil dependency, and high energy prices did energize many Americans on the urgent need to go green.

Decarbonizing our energy supply will continue to require innovation, funding, and sacrifice at every level of society. It will continue to be a long and arduous task. Nevertheless, all of us have an opportunity to contribute – to collaborate – in the global effort of going green. Green IT and green healthcare provide an excellent opportunity to save both energy and money!

Chapter Summary and Conclusions

From the above discussion on collaboration for green IT and green healthcare, the following conclusions can be reached:

- IT vendors have started to offer a significant set of integrated hardware, software, and services offerings to help customers improve their energy management initiatives.
- Industry organizations are establishing efficiency metrics at the server and data-center level to integrate facilities and IT resources.
- The EPA is establishing efficiency metrics at the server level as an extension to its Energy Star workstation metrics.

- Many IT companies are addressing the situation from end to end. At the server end through power management features and at the data-center ends through integrated IT/facilities modular solutions.
- The required collaboration for green IT is a part of the overall global collaboration required to solve the climate crisis.

Success for global green IT depends on the continued collaboration among groups within your company, technology vendors, data-center design and build businesses, energy utilities, governments, and organizations such as The Green Grid and the Uptime Institute. In short, almost everyone can collaborate on green IT, since almost everyone is a user of IT through PCs, the Internet, cell phones, etc.

Additional Reading

Koomey, Jonathan G. Ph.D. "Data Center Electricity Use" Lawrence Berkeley National Laboratory & Stanford University. Presented at the EPA stakeholder workshop at the Santa Clara Convention Center, February 16, 2007.

EU Code of Conduct. (2017). 2017 Best practice guidelines for the EU code of conduct on data centre energy efficiency. Retrieved from 2017 best practice guidelines for the EU code of conduct on data centre energy efficiency: https://e3p.jrc.ec.europa.eu/publications/2017-best-practice-guidelines-eu-code-conduct-data-centre-energy-efficiency.

Telemedicine and doctor-patient communication: An analytical survey of the literature, Edward Alan Miller, accessed on 27th October 2016 at the URL quoted below http://jtt.sagepub.com/content/7/1/1.short.

Electronic Information Security Policy in the context of Telehealth-http://www.unmc.edu/hipaa/_documents/telehealth-final.pdf. Accessed on 27th Oct 2016.

Telemedicine today and tomorrow: "Virtual" privacy is not enough by Christina M. Rackett accessed on 27th October 2016 at the URL: https://pdfs.semantic-scholar.org/85ea/9b153ec41d21a6daeeceac0a89599c518ceb.pdf.

Telemedicine privacy and security: Safeguarding protected health information and minimizing risks of disclosure, accessed on 27th October 2016 at the URLhttp://media.straffordpub.com/products/telemedicine-privacy-and-security-safeguarding-protected-health-information-and-minimizing-risks-of-disclosure-2015-08-13/presentation.pdf.

Privacy in a telemedicine environment, reference manual for health care professionals. Accessed on 27th Oct 2016 at the URL: https://support.otn.ca/sites/default/files/otn_reference_manual_for_health_care_professionals.pdf.

Chapter 8
The Need for Standard Healthcare and Hospital Energy Use and Carbon Footprint Metrics and the Triple Challenge

If You Can't Measure It, You Can't Manage It

—Quotation attributed to many, including W. E. Deming and
Peter Drucker

The Standard Performance Evaluation Corp. (SPEC) benchmark information has been used for years to compare servers from a power aspect. Here's the SPEC homepage: http://www.spec.org/. SPEC also has a benchmark to compare the power consumed by a server with its performance – a metric designed to aid users in boosting data-center efficiency. The EPA is pushing for metrics for all aspects of data-center energy use. In this chapter we'll look at an IT architecture that will help us meet the Triple Challenge for the healthcare industry: sustainability, privacy, and cloud-centric regulatory compliance.

The Triple Challenge for the Healthcare Industry: Sustainability, Privacy, and Cloud-Centric Regulatory Compliance

The healthcare sector is emerging as an important sector globally. Although the relative carbon footprint of healthcare is not overwhelmingly large compared to other industries, there are ensuing concerns for the sector arising from many quarters. Much has been written about "mMedicine," "mobility in health," "remote delivery of healthcare services," "eHealth," and "tele(health)care." The industry is looking to reap the benefits of emerging technologies such as mobile computing and cloud computing. These technologies are needed in order to grapple with ever growing operating cost problems. The challenge looms large as the healthcare sector is under heavy pressure due to regulatory compliance mainly for protecting the privacy of PHI (protected health information), as well as its contribution to sustainability through green practices in its IT infrastructure. Keeping such a wide mix in

© Springer International Publishing AG, part of Springer Nature 2018
N. S. Godbole, J. P. Lamb, *Making Healthcare Green*,
https://doi.org/10.1007/978-3-319-79069-5_8

mind, this chapter aims to build a framework for research investigation into the possible barriers for green IT practices in the healthcare sector as well as discussion about adoption of cloud-based application architectures in the healthcare sector.

Introduction

This chapter is organized into three sections. The first section is the background of the chapter; it addresses three key points: (1) the emerging importance of the global healthcare sector, (2) an overview of cloud computing and its importance for the healthcare sector, and (3) significance of the triple challenge point in a nutshell. In the second section, we present ideas for information technology architecture for the healthcare applications based on cloud computing and mobile technologies. A framework for research investigation is proposed in the third and last section of this chapter.

Globally, the healthcare (Wikepedia, n.d.) sector is emerging in importance and plays a key role in our economy (KPMG, 2011 October) (Doeksen & Schott, 2003) – see Fig. 8.1. Healthcare is not just another service industry; the healthcare sector is characterized by the fact that it brings together people and institutions involved in providing healthcare to those who require such services in times of need and stress. Healthcare is not only a complex industry but also a fragmented industry; it is composed of a number of sub-segments such as hospitals (a major sub-segment), pharmaceuticals, medical insurance agencies, medical equipment manufacturers, dealers and distributors, and diagnostics. Of these sub-segments, hospitals, by their very nature of operations, collect PHI (protected health information) (Protected Healthcare Information (PHI), n.d.)-related entities as well as treatment-specific confidential data.

The healthcare sector makes heavy use of technology including social media (KPMG, 2011 October). Technological advances have brought about revolution in the healthcare industry worldwide from modern testing techniques to improved surgical equipment, remote health monitoring technologies with the help of modern digital equipment, etc. There are a number of online healthcare portals.

Considering the complexity and scale of the healthcare industry, it is hardly a surprise that the sector has a large environmental footprint (UC Hospitals, 2009). For example, in the United States, it is estimated that healthcare facilities generate more than 5.9 million tons of waste annually. Hospitals contribute approximately 8% of the greenhouse gas emissions resulting from human activity. "Sustainability" in the healthcare sector can be classified into two types – (i) sustainability issues that are from non-IT areas and (ii) sustainability issues that are due to practices in the IT infrastructure of healthcare, mainly in hospitals. Of late, sustainability is getting integrated into healthcare investors' financial strategies (Becker, n.d.). Sustainable thinking for healthcare (World Economics Forum, 2013) (mainly for hospitals) comes from the "green-triad" – (a) green infrastructure, (b) green IT, and (c) green medical technology:

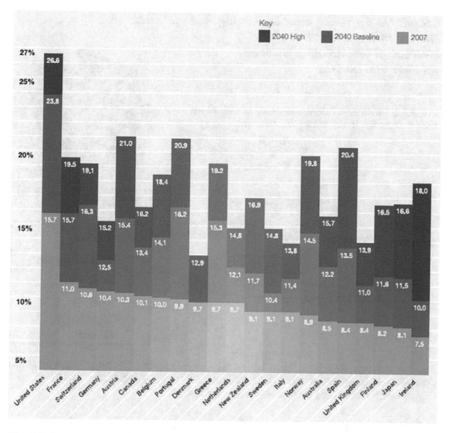

Fig. 8.1 Percentage of GDP for healthcare in OECD countries (projection). (Source: World Economic Forum, 2012)

- Cost savings through use of optimized and modernized heating, ventilation, and cooling (green infrastructure)
- Use of optimized illumination control systems for greater comfort and energy savings (green infrastructure)
- Deployment of customized solutions, services, products, and technologies to optimize the life-cycle performance for hospitals with a view to achieve maximum energy savings, performance, and sustainability without losing safety and comfort (green infrastructure)
- Less traveling by using modern communications systems, e.g., videoconferencing, telemedicine, etc. (green infrastructure)
- Green patient multimedia infotainment (green infrastructure)
- Desktop phones with reduced energy consumption (green infrastructure)
- Green transformation of healthcare data centers (Padhy, Patras, & Satapathy, 2012) through use of consolidation and virtualization technologies (green IT)

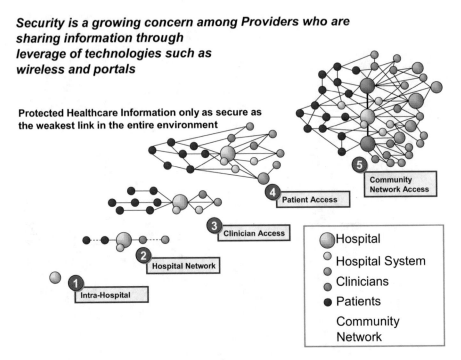

Fig. 8.2 Security: a major concern among healthcare providers. (Source: Impact of Cloud Computing on Healthcare Industry Green IT Practices, Nina S. Godbole, Dr. John Lamb, AoT Regional Face to Face IBM Somers, New York, April 24, 2013)

- Improving data-center energy efficiency (Lamb, 2009) – file server storage reduction and usage of groundwater for air cooling in DC (green IT)
- Green IC architectures – desktop and server virtualization (green IT)
- Energy-efficient medical technologies (green medical technologies)
- Making operation of medical technologies environmentally friendly (green medical technologies)
- Healthcare waste and eWaste (electronic and electrical equipment) (Godbole, 2011a, 2011b) disposal through reduce-reuse-recycle principle (green infrastructure)

Much of the focus today is on the greening of data centers and the greening of IT in general (Lamb, 2009). As we've discussed, healthcare sustainability is very much involved with green data centers and green IT. The complex extended chain of the healthcare network for hospitals is depicted in Fig. 8.2. In such a complex of data interconnections for hospitals, IT security and the potential for "data breaches" are of great importance. So, when healthcare organizations strive for IT energy efficiency, they also need be concerned with IT security.

Although healthcare organizations use a great variety of IT applications and infrastructures, patients may not always see that reflected into the costs to the patients as IT is not healthcare organizations' primary activity. Moreover, due to the rapid obsolescence of information technology, the IT systems and applications used in healthcare need to be updated frequently. The cost of IT systems in healthcare services is quite high, and often, many of healthcare organizations pass this cost on to their patients. This is how the healthcare delivery system can become expensive from an end user's perspective.

The five most popular models for cloud deployment are SaaS (Software as a Service), PaaS (Platform as a Service), IaaS (Infrastructure as a Service), DaaS (Desktop as a Service), and BPaaS (Business Process as a Service).

The four main types of cloud computing architectures noted in the healthcare sector are:

- External/public: Shared computing resources maintained by an off-site third-party provider that offers pay-per-use access to data, applications, infrastructure, etc. The healthcare application developers and consumers are fully responsible for protecting patients' security and privacy.
- Internal/private: The cloud infrastructure is operated solely for a healthcare delivery organization. Dedicated computer resources provided by an off-site third-party or use of cloud technologies on a private internal network that is managed and maintained by in-house IT staff.
- Community: The cloud infrastructure is shared by several healthcare delivery organizations and supports a specific community that has shared concerns (e.g., security requirements policy, mission and compliance considerations). It is typically managed by the third party or the healthcare delivery organizations and may exist on or off premise.
- Hybrid: Multiple mixed-use public/private clouds, integrating on-site IT cloud infrastructure with third-party provider services.

With the rising trends of collaborative healthcare and the economic pressure for remote delivery of healthcare services, we can imagine that most of the patient data being exchanged in the healthcare sector is considered as "sensitive." The world over, security and privacy of PHI (World Economics Forum, 2013) is a recognized issue, and there are data privacy protection guidelines for the healthcare sector, for example, HIPAA in the United States.

Sustainability is another major aspect of healthcare sector operations worldwide, in view of the impact of CFCs (chlorofluorocarbons) on our environment and carbon footprint of healthcare. Table 8.1 shows the typical functional areas in hospital in terms of their carbon emissions.

Medical records of a patient may refer to PHR, EMR, and HER – see Fig. 8.3. Personalized privacy requirements would impact the degree of overlap among the three types of information mentioned in the figure.

There have been a number of studies on green practices in the healthcare sector (Godbole, 2011a, 2011b) and barriers to green practices (World Economics Forum, 2013) in the healthcare waste sector. There is indeed a need to rethink healthcare systems. To design more sustainable health systems, advantage must be taken of demand-side opportunities.

Table 8.1 Hospital: functional area-wise energy consumptions and carbon emissions

Functional area of use	Energy consumption	Carbon emissions
Ward	1350	293
Surgery areas	844	268
Back area	619	176
Consulting areas (consulting rooms, etc.)	510	166
Administration and office blocks	474	154
Corridors (24 h)	349	95

The carbon emissions were calculated using the following Scope 3 emission factors: 1 kWh electricity = 1.35 kg CO2-e and 1 GJ of gas = 55.7 kg CO2-e

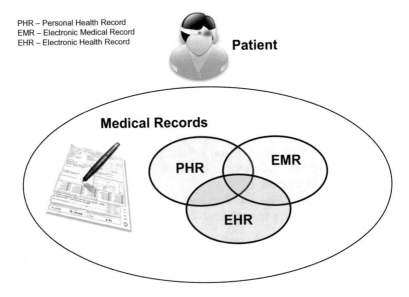

Fig. 8.3 Healthcare data privacy terms

The third dimension of the challenge to healthcare sector comes in the form of regulatory requirements. As mentioned before, in healthcare, cloud computing can support a wide range of organizational structures and business services, such as electronic health records, e-prescribing, practice management, computerized physician order entry, billing and administration, etc. Cloud technologies are of particular interest to healthcare industry in the United States, where health systems are provided with the mandate to qualify for more than $20 billion in US government financial incentives under the HITECH Act (Health Information Technology for Economic and Clinical Health). In this regard, the challenge lies in meeting deadlines for the meaningful use of certified electronic health record (EHR) technology. Small and large providers are finding that cloud-based EHRs can be implemented quickly and cost-effectively.

Patient records are sent in digital form through the extended healthcare network, giving an opportunity for pilferage of sensitive information (PHI). Regulating the cloud and audit considerations is one of the major challenges from the PHI perspective, for the healthcare industry. When the data is in a public IT infrastructure, opportunities for stealing sensitive data exist. For example, a 2010 study concluded that breaches cost the healthcare industry about $6 billion per year. Another 2011 study indicates that the total economic impact of medical identity theft is $30.9 billion annually, up from $28.6 billion in 2010.

IT Architecture for Healthcare Based on the Triple Challenge

Examples of Cloud Computing Scenarios in Healthcare

Consider the following scenarios as examples of cloud-computing-based healthcare scenarios in the years to come:

Scenario 1: a hospital, due to its limited IT budget, is finding it difficult to process and store in-house, voluminous data consisting of medical images. The hospital decides to use the IT facilities available from cloud service providers to outsource the processing and storage of medical images.

Scenario 2: a Spaniard, living in Madrid has been sent to a training program in the United States. While commuting from the airport to the hotel, his taxi is hit by a car approaching from the opposite direction. The person becomes unconscious and is in a need for immediate treatment including a blood infusion and several life-saving medicines. The doctor quickly scans a tag of one of his electronic devices (iPhone, tablet, iPad, Internet-enabled watch, etc.). The tag connects the doctor to an online database available in the cloud that identifies the patient and instantly retrieves relevant information on patient's blood type and allergies to medication. Fortunately, a critically needed medicine is noted in these records, enabling the patient to be saved.

Security and Privacy Requirements in Healthcare

Given the extended nature of healthcare service networks and as explained earlier, there is relationship among EMR, HER, and PHR. Patients' multiple EMRs may be stored with several healthcare delivery organizations, depending on the entity that provides the service. This could be in addition to the patients' PHRs and EHRs. Let us take a scenario after the patient visits healthcare practitioners in hospital or other healthcare delivery agencies, some of EHRs are obtained from various EMR systems, while some EHRs, for example, historical health information, could be held by the patient or by the patient's family members. Under this scenario, there are key challenges from security and privacy perspective:

- The first challenge is about access management, i.e., how to manage and control the access of the EMR data in the EHR system, as accessing EMR data is typically controlled through authorization models.
- The second challenge comes due to the fact that a patient may not want to divulge some of his sensitive health information in his/her EHRs (either to his/her family members or to all or some healthcare providers. Thus, the requirement of preserving privacy in access to EHRs needs to be addressed.
- Thirdly, the authenticity of EHR data also needs to be addressed with respect to both content authentication and source verifiability.

IT Infrastructure Architecture in Healthcare Systems

With the rapid growth of healthcare industry, the reliance on IT increases dramatically. Newer strategies need to be used in IT infrastructure for healthcare industry to be able to cater to the need for reducing the cost of healthcare delivery and improving the quality of healthcare delivery and to extend the reach of healthcare services, especially in the rural areas where use of cloud-computing-based applications has a great potential (Padhy et al., 2012), for example, making medical expertise available to rural people through a collaborative use of cloud-based medical databases of past history and patterns for use of patient treatment. There is also a need for innovative IT solutions to support the ability (i) to survive the inevitable changes in healthcare industry and (ii) to enable a competitive advantage for the industry. Given the rising costs of healthcare service operations and the need to improve the reach of healthcare services, mobile medicine is becoming an imperative. In healthcare IT infrastructure, both the Internet and wireless sensor networks are deployed to achieve mobility in monitoring patient conditions. For example, sensors are attached to patients for monitoring and measuring their vital signs, e.g., a patient's heart rate or body temperature. Such measurements are sent periodically to the medical staff. The challenge in IT infrastructure design for the healthcare lies in managing the trade-off between the level of security provided and the resources consumed. Mobile access can provide significant benefits such as easy access to medical information and energy savings. However, security and privacy issues under mobile access need to be carefully considered.

There are a few considerations important from regulatory requirement perspective. For example, patient records hold sensitive information (PHI) and may reside on numerous systems deployed on the cloud. Therefore, healthcare organizations need to view patient record in the cloud with data privacy considerations. In addition, healthcare entities may need visibility into their own applications given the more stringent requirements in HIPAA/HITECH. Business associates may also want visibility of provider environments as well. To gain visibility into the use of

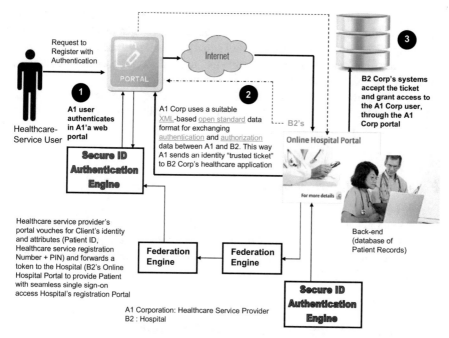

Fig. 8.4 Federated identity used in healthcare system

cloud applications, it is very critical to monitor access and data flow from the medical organization through the cloud. When using public clouds infrastructure, there are multiple audit considerations to support regulatory requirement such as SAAS 70: physical, cyber, operational, and policy.

One of the functions available to these healthcare service portal users is a service specification database for all medical services arranged by A1 (the healthcare service provider) and its channel partners (pharmacies, hospitals, medical equipment retail and bulk sellers, etc.). This database is hosted and managed by B2 Corporation, and access is restricted according to service agreements with B2's affiliates, such as A1. The interactions, under the FIM (federated identity management) scenario, are depicted in Fig. 8.4.

The effect of this architecture is that when an A1 customer logs in to the portal and clicks the "Med-services Specs" link, the B2 Corporation database search application is served up via the A1 portal. A1's customers do not need to maintain a separate patient ID/user ID and password within B2's environment, and neither company has to synchronize passwords, IDs, or profiles. Behind the scenes, federated identity systems at both corporations (A1 and B2) transparently manage the steps required to make the above scenario possible; the technical description of this is beyond the scope of this chapter.

The healthcare service delivery cost issue was mentioned in an earlier section of this chapter. A cloud-based architecture would be especially appropriate for a rural healthcare center. In that architecture, a cloud controller would manage physical resources, monitor the physical machines, place virtual machines, and allocate storage. The controller would respond to new requests or changes in workload by provisioning new virtual machines and allocating physical resources. In reference to the FIM use mentioned (see Fig. 8.4), note that a cloud service provider would need to use an authentication and authorization mechanism.

"Authentication" means that each user has an identity which can be trusted as genuine. "Authorization" means that each resource will have a set of users and groups that can access it.

Proposed Framework for Research Investigation

In this section, we propose a research investigation approach for undertaking a case study-based method for studying the aspects mentioned in the previous sections. Considering the three areas addressed in this chapter, we propose that a research investigation could proceed as outlined below.

Sampling and Sampling Frame Considerations

In an investigation along the triple challenge points – (i) sustainability, (ii) privacy, and (iii) cloud-centric regulatory compliance – we envisage that drawing the sampling frame for the research would be the first challenge. This is because, as mentioned in the beginning of the chapter, "healthcare" is a very large and also a fragmented industry. Although a sample frame can be drawn on "hospital directory" of a region, a researcher ought to pay attention to possible sampling methods and sampling errors. Depending on the size of the population, a stratified sample could work. To narrow down the scope of the investigation, even if the research were to focus only on the "hospital" component of healthcare industry, "classification" of hospitals would be a key consideration (e.g., based on number of beds, types of specialties offered, etc.). Classifying hospitals as primary, secondary, or tertiary institutions is also common. Hospitals can also be classified based on "ownership," for example, government, semi-governments, voluntary agencies, private commercial, private charitable trusts, etc.

Relating Sustainability to Cost in Healthcare Sector

The next challenge would be finding the link between "sustainability" and cost of healthcare delivery. As mentioned earlier, IT is not healthcare organizations' primary activity. Therefore, we envisage that most stakeholders in healthcare would not consider "greening" of healthcare as their immediate goal.

Risks to Healthcare Sector in Adoption of Cloud-Computing-Based Solutions

While the cloud computing scenarios presented for healthcare have a great appeal, risks to sensitive data are high. Specifically, with cloud computing healthcare, the sector faces the following risks:

- Breach of legitimate patient records through medical identity theft
- Financial/billing fraud
- Unauthorized access to PHI
- Privacy loss

Thus, the adoption of cloud computing technologies brings up PHI protection and access monitoring issues. In this regard, careful example of FIM (federated identify management) and related mechanism would need to be carefully examined.

Mixed Method Research

A mixed method research design (Brannen, n.d.) is a procedure for collecting, analyzing, and "mixing" both quantitative and qualitative research and methods in a single study to understand a research problem. Normally, mixed method is used when research questions involve exploring the meaning of a "research construct" or a phenomenon from more than one perspective. This is applicable given that there are differences between the two research areas; the most important to note is that while "quantitative" research "tests" theory, "qualitative" research "develops" theory. Equally important, qualitative research is "interpretive" rather than measurement oriented.

Mixed method studies make use of quantitative and qualitative methods at the start of the research process. There is a considerable literature available to suggest "mixed methods" and "triangulation" (Morgan) as a possible and acceptable approach for "qualitative" research rather than pure "quantitative" research. Quantitative research is concerned with numbers, objective hard data. In quantitative research, the sample size for a survey is calculated with the use of formulas, to determine how large a sample size will be needed from a given population in order to achieve findings with an acceptable degree of accuracy. It is a common practice to seek sample sizes which yield findings with at least a 95% confidence interval. In contrast to this, qualitative research is concerned with collecting, analyzing, and interpreting data by observing what people do and say. While "quantitative research" refers to counts and measures of things, "qualitative research" refers to the meanings, concepts, definitions, characteristics, metaphors, symbols, and descriptions of things. Unlike quantitative research, qualitative research is much more "subjective" and uses more unconventional methods of collecting information, mainly individual, in-depth interviews and focus groups. The nature of qualitative research is exploratory and open-ended. Small numbers of people are interviewed in-depth and/or a relatively small number of focus groups are conducted. To present an analogy, whereas "quantitative" research is about "counting the beans," "qualitative research" is about providing information as to "which beans are worth counting"!

Literature (Brannen, n.d.) (Morgan) suggests that there are five major purposes or rationales for conducting mixed method research: (i) triangulation (i.e., looking for convergence and corroboration of results from various methods and designs studying the same phenomenon); (ii) complementarity (i.e., looking for explanation, enhancement, illustration, and clarification of the results from one method with results from the other method); (iii) initiation (i.e., making discovery of paradoxes and contradictions that lead to a rephrasing of the research question); (iv) development (i.e., using the findings from one method to help inform the other method); and (v) expansion (i.e., seeking to expand the breadth and range of research by using different methods for different inquiry components).

Given the type of challenges explain in previous chapters and given the nature of healthcare industry, we strongly believe that a qualitative approach would suit this type of investigation. There is a considerable scope for using "mixed method" research. There are three types of mixed method design that we recommend for taking up investigation of the issues involved in the triple challenge for healthcare industry – (i) triangulation mixed method design using both "qualitative" and "quantitative" data to interpret data, (ii) explanatory mixed method design conducting a follow-up qualitative study after a quantitative study, and (iii) exploratory mixed method design – building a quantitative study on the results from a qualitative study.

Chapter Summary and Conclusions

In this chapter, we discussed the need for standard healthcare and hospital energy use and carbon footprint metrics. As discussed in this chapter, emerging technologies such as mobile computing and cloud computing hold great promise to reduce costs and significantly improve services in the healthcare industry. However, healthcare industry challenges are continuing to grow in the areas of regulatory compliance for protecting patient privacy. Also, the continued significant use of electric energy for the IT infrastructure used to support the healthcare industry has increased pressure on the industry to support green IT initiatives and overall sustainability.

This chapter's proposed IT architecture for healthcare based on the triple challenge of sustainability, privacy, and cloud computing holds great promise. However, continued research to validate the proposed IT architecture is needed. The proposed framework for research given in the last section of this chapter will help clarify the overall direction needed.

Additional Reading

Al khuwaildi, H. A.. (2010). Green techniques in the healthcare system, a thesis in partial fulfilment of the requirements for the Degree Master of Science. Jordan: Jordan University of Science and Technology.

Godbole, N. (2009). Information systems security: Security management, metrics, frameworks and best practices, chapter 29 and chapter 30, Wiley India Pvt. Ltd, ISBN 13: 978-81-265-1692-6.

Muduli, K., & Barve A. (2012, August). Barriers to green practices in health care waste sector: An Indian perspective. *International Journal of Environmental Science and Development, 3*(4), 393–399.

Optimizing Healthcare Information Sharing, Data Privacy and Compliance – Titus Lab White Paper – http://www.handddatagovernance.com/whitepapers/Whitepaper_Healthcare_Titus_Labs_Information_Sharing.pdf.

Personal Health Record (PHR) Wikipedia definition at – http://en.wikipedia.org/wiki/Personal_health_record. Another definition of PHR at – http://searchcompliance.techtarget.com/definition/personal-health-record-PHR.

Rabi Prasad Padh, Manas Ranjan Patra, Suresh Chandra Satapathy. Design and implementation of a cloud based Rural Healthcare Information System Model, Oracle Corporation, Bangalore, India.

The PHR tool – http://www.myphr.com/startaphr/what_is_a_phr.aspx.

Chapter 9
Green Healthcare Economics: The Financial Incentives

Pocketbook environmentalism is powerful. If marketers can help consumers understand the hidden costs in products and services that are not environmentally friendly, they can grab consumers' attention.

Joel Makower – Author of "Strategies for the Green Economy"

The case studies described in this chapter include examples of how green healthcare can not only help "save the planet" from global warming but also save the healthcare system significant money. So, green healthcare is a true "win-win" opportunity. Not only do you help society in the battle to combat climate change, but you save money doing it! It should be noted that the case studies in Chaps. 10 and 11 also give examples of how green healthcare saves money, but the focus in Chap. 10 is green healthcare opportunities in US hospitals, and the focus in Chap. 11 is how green healthcare is being practiced worldwide. The significant value in looking at green healthcare case studies, for both large and small healthcare facilities, should be to take the opportunity to take advantage of the case study lessons learned.

Examples of How Greening the Medical Center Saves Money

The following sections give examples of the different ways that green healthcare can save money.

Hospitals Save Money by "Going Green"

Much of the following information was taken (with permission) from the following article (http://www.healthcarefinancenews.com/news/hospitals-save-money-going-green).

© Springer International Publishing AG, part of Springer Nature 2018
N. S. Godbole, J. P. Lamb, *Making Healthcare Green*,
https://doi.org/10.1007/978-3-319-79069-5_9

Achieving LEED Certification Is Becoming a Cost-Effective Goal

While it may cost slightly more initially, many hospital and healthcare facilities around the country are finding it financially beneficial to receive LEED certifications by the US Green Building Council (USGBC) for sustainable designs, green building materials, and energy-efficient systems.

LEED stands for Leadership in Energy and Environmental Design and is the accepted benchmark for sustainable building practices in renovating existing facilities or building new ones. Using a stringent rating system, the USGBC certifies projects Silver, Gold, or Platinum.

Example of LEED Building Monetary Savings

In February, 2016, the Katz Women's Hospital at North Shore University Hospital in Manhasset, N.Y., which is part of the North Shore-LIJ Health System, became the first hospital project in New York to be awarded the Platinum LEED certification. The 73-bed Katz Women's Hospital at North Shore features single maternity rooms that take advantage of natural light, private single rooms for mothers, a well-baby newborn nursery, and work areas. With the Platinum rating, on average, the Katz Women's Hospital uses 18.6% less energy and 51% less water, and 100 % of the electrical power consumed in its first 2 years of operation will have been generated using renewable energy from wind power in Texas.

Case Study: Applying Resource Modeling to IT Problems

Much of the following case study information was taken (with permission) from the following article by Pranali Humne (https://www.isixsigma.com/implementation/case-studies/case-study-applying-resource-modeling-to-it-problem/).

The iSixSigma organization has created many case studies showing how the Lean Six Sigma process can improve operations and save money. The case study given in the URL above gives details on the use of Six Sigma resource modeling to IT problems. The information given below lists the cost savings given in the case study article. Access the URL for details on the case study.

Resource utilization is managed using models. These models are created based on multiple influencing variables or factors. The goal is to balance resources – including people and skill levels – to achieve higher service levels, lower costs, and other benefits.

Widely used across industries, resource modeling functions as an efficient decision-making tool. Some successful applications include the following:

- A retail organization recorded an annual savings of $52 million.
- A consulting firm increased its share of business by 35%.
- A call center raised productivity to 98%.

This article uses a hypothetical case within the field of information technology (IT) to show the mechanics for building these models. The reduction of time to provide a customer with information is the key process output variable, but there are others. The input variables will be discussed using an input-process-output (IPO) diagram and various analytical methods.

What Can Resource Modeling Do?

Deciding on the skill types for human resources as well as how many employees are needed during the initial setup for a recently migrated process (moving a process from one computing platform to another) is a big challenge. This is especially true for resourcing decisions needed to create steady-state operations. Lean Six Sigma tools and methods are useful for the stabilization and improvement of performance metrics, such as the time needed to respond to a customer issue, the time to resolve an issue, and compliance with service-level agreements.

Resource modeling helps answer key questions such as:

- Which resource should be deployed?
- What are the constraints and the expected outcomes?
- Are the influencing factors interrelated?
- How should human resources be based – geography, skills, or other factors?
- How can maximum productivity and efficiency be ensured?
- What is the optimum number of employees required to ensure timely response to customer inquiries?
- How can cost-effectiveness be maximized?

Using Lean Six Sigma (LSS) for resource modeling ensures a bias-free, precise, and accurate outcome. It also helps create models based on multiple combinations resource estimation, feasibility, control, and simulation.

Example: Server Migration

The migration of multiple servers across the world is used as an example for resource modeling. The servers are the inputs or independent variables. The output time, in seconds, is the response variable. The resources to be modeled are employees. See the URL above for details of the server migration model.

Summary

In summary, resource modeling helps to plan and determine the number of human resources required (here for an IT support project) to operate smoothly. It enables achieving the desired response rate and other targets efficiently. It also helps to lower operational costs and make key decisions about staffing levels in terms of numbers and skill levels. All help achieve an organization's LSS goals.

Pointers to Other Chapters for Additional Details on Case Study Lessons Learned

This case study emphasized the need for collaboration, green IT, etc. Here are pointers to the chapters that give additional detailed information on those topics.

- Collaboration. Chapter 7
- Green IT. Chapters 5 and 7
- Network Design and Data-Center Efficiency. Chapters 3 and 5
- Building Automation Systems. Chapters 3 and 5
- Six Sigma Process. Chapters 2, 4, and 10

Stanford University Medical Center, California, USA

The Path of Least Resistance: Is There a Better Route?

Much of the following case study information was taken (with permission) from the following article (https://www.isixsigma.com/implementation/case-studies/path-least-resistance-there-better-route/).

The iSixSigma organization has created many case studies showing how the Lean Six Sigma process can improve operations and save money. The case study given in the URL above gives details on the use of the Lean Six Sigma process to delivering quality patient care with the highest level of efficiency. The information given below lists the cost savings given in the case study article. Access the URL for additional details on the case study.

Driven by nationwide technologist shortages, an industry-wide focus on quality and rising consumer demand, healthcare is feeling the pressure to deliver more with less. Given this challenging environment, some organizations have come to the conclusion that taking the path of least resistance may not be the best approach.

For many, delivering quality patient care is nonnegotiable, and the ability to function at the highest level of efficiency while doing so is seen as a necessity, not a choice. Realizing and responding to this quandary a few years ago, California's

Stanford University Medical Center decided to aggressively work toward a solution and overcome resistance by approaching it with evidence.

Developing One Clear Vision

In May 2000, the radiology department at Stanford University Medical Center embarked on a 5-year journey toward complete digitization, partnering with GE Medical Systems to make this transition. While the end goal was known, there was much less certainty about the steps involved along the way. Initiating any large-scale change within an academic medical institution presents a unique set of challenges, and a sweeping technological transformation clearly impacts traditional workflow and inherent cultural barriers.

To smooth the transition, Stanford worked with GE's team to implement Six Sigma process improvement methodologies and related change management techniques, making the steps toward a digital environment much clearer and easier to follow. The methodical and evidence-based framework of Six Sigma significantly organized the process of "going digital" by breaking it into manageable projects with clear objectives.

Change management techniques used in conjunction with Six Sigma were equally instrumental in helping the team at Stanford gain acceptance among staff for the changes introduced into the system. Combining technical and cultural strategies helped to pave the way during Stanford's journey toward digitization and illustrated the importance of addressing the human side of the equation through workflow and behavior modification.

Identifying and Overcoming Roadblocks

As the process began, key stakeholders within Stanford's radiology department sought to identify and focus on the most significant issues impacting productivity within the radiology department. Appropriate selection and scoping of projects on the front end is a critical aspect of the Six Sigma approach. Stanford identified five key areas where improvement could be made that would have a major impact on the organization: MR outpatient throughput, CT inpatient throughput, CT outpatient throughput, report turnaround time, and Lucile Packard Children's Hospital CR/Ortho throughput and digitization. These five areas became the initial Six Sigma projects at Stanford, and the CT project is presented below.

To represent key stakeholders, projects were assigned team members comprised of management, staff, and radiologists. Over a period of 6–12 months, each project team typically met two to three times per month to review progress and plan next steps.

Resource constraints were taken into consideration as Stanford began the Six Sigma initiative. Although labor-intensive, collecting radiology data manually is often the best way to obtain the level of detail required, unless there is a robust RIS in place with solid data integrity. To gather the necessary information without unduly impacting staff and workflow at Stanford, the consultants working onsite handled the actual observation and recording of data.

CT = "Computed tomography"

A *CT scan* makes use of computer-processed combinations of many X-ray measurements taken from different angles to produce cross-sectional (tomographic) images (virtual "slices") of specific areas of a scanned object, allowing the user to see inside the object without cutting. Other terms include computed axial tomography (CAT scan) and computer-aided tomography.

Gaining Momentum in CT

Kicking off in September 2001, the CT Six Sigma project focused on increasing throughput and decreasing the outpatient-scheduling backlog. In the initial phases of the project, scheduling backlog was reported at 16 days. Even more frustrating was the increasing number of patients who missed appointments, which is not unusual when patients and referring physicians are waiting more than 2 weeks for a CT appointment.

One of the first steps in understanding and improving capacity was to collect throughput data at Stanford's Blake Wilbur Outpatient Center. There were numerous factors affecting the data gathering process, and in this situation the onsite consultant observed and recorded the times involved with providing CT services and moving patients through the system.

Prior to the project, CT appointments were scheduled every 30 min during the day shift and every 45 min on the evening shift. The staffing pattern on the day shift was 1.0 FTE CT technologist and 1.0 FTE technical assistant. The evening shift was staffed with 1.0 CT technologist, but no technical assistant, which was the rationale for the longer appointment times. Radiology nurses handle the IV access and cover both MR and CT at the outpatient facility.

The CT throughput data was analyzed and the results were presented to the team for recommendations. The data revealed that the mean CT exam in-room time was 16.9 min with a standard deviation of 7.8 min representing 46% of the mean. It was obvious that the evening shift in-room times were comparable to the day shift and that extended appointments were not necessary. One of the first changes to be implemented was to convert the 45-min appointment times to 30 min, resulting in a gain of 28 potential appointments per week. This adjustment was the first of numerous design solutions that were implemented in CT at the Blake Wilbur Outpatient Center.

Pointers to Other Chapters for Additional Details on Case Study Lessons Learned

This case study emphasized the need for collaboration, green IT, etc. Here are pointers to the chapters that give additional detailed information on those topics.

- Collaboration in General. Chapter 7
- Green IT. Chapters 5 and 7
- Network Design and Data-Center Efficiency. Chapters 3 and 5
- Six Sigma Process. Chapters 2, 4, and 10

Chapter Summary and Conclusions

A significant motivation for implementing green healthcare should be the fact that green healthcare can save healthcare providers a lot of money. The case studies described in this chapter give examples of the cost savings. Chapters 10 and 11 also give case study examples where significant money is saved, but the focus for the case studies in Chaps. 10 and 11 is on improving healthcare efficiency and effectiveness by going green, for healthcare facilities in the U.S. and worldwide.

As stated at the beginning of this chapter, the significant value in looking at green healthcare case studies, for both large and small healthcare facilities, should be to take the opportunity to take advantage of the case study lessons learned.

Additional Reading

Muduli, K., & Barve, A. (2012, August). Barriers to green practices in health care waste sector: An Indian perspective. *International Journal of Environmental Science and Development, 3*(4), 393–399.

Chapter 10
Case Studies for Green Healthcare at US Medical Colleges and Large and Small US Hospitals

Ah, to build, to build! That is the noblest of all the arts.

—Henry Wadsworth Longfellow,

from the poem, "Michael Angelo" (pt. I, II, l.54)

The case studies described in this chapter include "lessons learned" for medical colleges and large and small hospitals in the United States. Universities offer a rich environment for green healthcare innovations and best practices. Not only do they have the administrative systems common to any large healthcare facility, but with their university research departments, they have many groups of researchers in varied fields who can help improve the university healthcare energy efficiency and carbon footprint.

Of course, large hospitals (including university hospitals) typically have many more resources for making the hospital green. However, the significant value in looking at green healthcare case studies, for both large and small healthcare facilities, should be to take the opportunity to take advantage of the case study lessons learned.

Examples of Greening the Medical Center

University of California, San Francisco, Office of Sustainability

Much of the following green healthcare information for this chapter was taken (with permission) from the following web site for the University of California, San Francisco (UCSF), Office of Sustainability. http://livinggreen.ucsf.edu/greening_the_medical_center/healthcare_sustainability_news1

The UCSF Office of Sustainability web site contains many articles on how to make healthcare green.

© Springer International Publishing AG, part of Springer Nature 2018
N. S. Godbole, J. P. Lamb, *Making Healthcare Green*,
https://doi.org/10.1007/978-3-319-79069-5_10

The University of California, San Francisco, is the medical college for the ten-campus University of California system. The University of California, San Francisco, Office of Sustainability, encourages other medical centers to reference their web site and the Greening the Medical Center section as they want to share Living Green Clinic/unit certification forms, Climate Changes Health posters, and other best practice materials developed by the UCSF. Two of the USCF sustainability stories from 2017 are given below. These sustainability stories include many "lessons learned" that can be valuable to other medical centers in their quest to become green.

UCSF Labs: Five Ways to Use Less Power and Save More Lives

The author of this UCSF sustainability story is Deborah Fleischer, Green Impact, September 2017.

See URL: http://sustainability.ucsf.edu/3.691.

UCSF has over 2.6 million gross square feet of buildings focused on research. These labs are developing life-saving cures to some of the most challenging diseases and debilitating injuries, but the electricity that powers our life-saving research has a shadow side – a carbon footprint.

A new poster campaign launched by the Office of Sustainability has a clear call to action: reduce wasted energy by turning off lab equipment and monitors during non-occupied hours. Some simple, small actions on your part can collectively make a difference:

- Turn off unused equipment.
- Set ULTs at −70 °C. ULTs are ultralow temperature freezers for lab and medical storage.
- Buy Energy Star equipment.
- Use timers on equipment.
- Get LivingGreen Certified.

Energy Use Intensity in Labs

Labs consume significantly more energy per square foot than the average building due to specialized equipment, such as laboratory fume hoods, −80° freezers, and other research equipment. Collectively UCSF's labs are very energy intensive, using over 500,000,000 kBTUs in FY16, which is equivalent to driving 21,753 cars for a year. This carbon footprint has an impact on the health of UCSF's patients, ranging from an increase in asthma to more premature births to increased vulnerability for youth and the elderly from extreme weather events.

One way to compare energy use across labs is to calculate energy intensity (EUI) in kBTU/square foot (sq. ft.). The EUI expresses a building's energy use as a function of its size or other characteristics; it is calculated by dividing the total energy consumed by the building in 1 year (measured in kBTU) by the total gross floor area of the building.

The US national average EUI for a lab is 370, with a goal of 111 for newly constructed labs. UCSF has set two different goals for its existing labs – at the Parnassus UCSF facility, the goal is to reach 200 kBTU/sq. ft., and at the Mission Bay facility, the goal is to reach 175 kBTU/sq. ft.; the goal at Parnassus is higher to take into account the older age of the buildings, which are less energy efficient.

At the Mission Bay UCSF facility, which are on average newer, more energy-efficient buildings, EUI ranges from a low of 190 at CVRI to a high of 397 at Sandler Neurosciences Center. At Parnassus, where the buildings are older, EUI ranges from a low of 224 at Health Sciences, which includes both HSE and HSW, to a high of 273 at PSSRB.

Call to Action

If all labs incorporate these simple actions, the UCSF community can help save more lives by reducing UCSF's greenhouse gas emissions that contribute to health-related climate change impacts:

- Turn off equipment.
- Set ULTs at −70 °C.
- Buy Energy Star equipment.
- Use timers on equipment.
- Get LivingGreen Certified.

1. *Not Using It? Turn It Off*

This first one is obvious, but a friendly reminder always helps. If a piece of equipment is not being used, turn it off. This applies to lights, centrifuges, shakers, computer monitors, and fume hoods. Consider trying the Adopt-a-Spot campaign, which engages green champions to take ownership of specific pieces of equipment. In a pilot, this program reduced energy use at labs by 8–9%. Another easy way to reduce energy and waste in the lab is to identify equipment that does not need to run 24/7 and use a timer to turn the items off in the evening and back on in the morning. Items such as heating blocks and mixers can easily be plugged into one power strip and programed to automatically turn off at the end of the day.

When left plugged in, our electronic gadgets and basic appliances still use what's called phantom or vampire energy – even when they are turned off or in sleep mode. For example, a plugged-in cell phone charger consumes energy even when it's not charging your phone. There is a simple solution: unplug it when not in use. The use of a power strip can make this easy to do for multiple devices. Using a "smart" power strip that automatically shuts off when devices are inactive makes this even easier.

2. *Chill Up: −70 °C Is the New −80 °C*

According to My Green Lab, chilling up your ultralow freezer from −80 degree Celsius (C) to −70 °C has two major benefits: it can reduce energy consumption by 30% and in doing so it can prolong the life of your freezer. This means less down time and less chance that your samples will be compromised.

3. *Buy Energy Star and EPEAT*

If you have procurement responsibilities, for all new equipment, look for energy star, EPEAT, or energy-efficient options. Go to https://www.energystar.gov/products/spec/laboratory_grade_refrigerators_and_freezers_specification_version_1_0_pd to see Energy Star lab-grade freezer and refrigerator options. For desktops, laptops, imaging equipment, mobile phones, and televisions; EPEAT, the leading global ecolabel for the IT sector, provides institutional purchasers an easy way to identify and compare high-performance, more-sustainable products.

If your −80° freezer is over 10 years old and/or has mechanical issues that cause excessive consumption of energy, such as overactive compressors or poorly sealed doors, consider replacing it with a new, energy-efficient model. And for those appliances and instruments for which Energy Star is not available, there may be utility-company-sponsored rebates to help offset the cost of purchasing energy-efficient equipment; check with the Office of Sustainability for more information on rebates.

If energy-efficient equipment is not available, be clear with companies that you want such products. When Thermo Fisher inquired about the UCSF rebate program to encourage the purchase of energy-efficient ULTs, we asked that they offer a 50% more energy-efficient freezer. Within a year, they met the challenge.

"Reducing energy demand in labs is one of the most significant ways to reduce our carbon emissions. When customers demand energy efficient equipment, manufacturers will rise to the occasion," explained Gail Lee, UCSF's Sustainability Director.

4. *Use Timers on Equipment*

Another easy way to reduce energy and waste in the lab is to identify equipment that does not need to run 24/7 and use a timer to turn the items off in the evening and back on in the morning. Items such as heating blocks and mixers can easily be plugged into one power strip and programed to automatically turn off at the end of the day.

5. *Get LivingGreen Certified*

A new UC-wide green lab policy is encouraging all the campuses to amp up the greening of their labs by having three labs per year be green lab certified. At UCSF, labs can already get certified through the LivingGreen Certification Program. The four-page checklist provides a wealth of possibilities for reducing energy, as well as for tackling waste reduction, conserving water, and reducing toxic components. Even if you aren't ready to tackle getting certified, the checklist is a useful list of green lab best practices.

To learn more go to http://livinggreen.ucsf.edu/greenlabs and 2017 S-Lab Awards Results.

Carbon Neutrality Fellows Incorporate Environmental Topics onto Education

The author of this UCSF sustainability story is Kailyn Klotz, Sustainability Fellow, 2017.

See URL: http://sustainability.ucsf.edu/3.701.

"Climate change is something that our profession is going to have to confront in a really big way, and that attitude is not reflected in the curriculum right now," said UCSF medical student, Carolyn Rennels.

Through the Carbon Neutrality Initiative (CNI), second-year medical students, Gabriela Weigel and Carolyn Rennels, have spent much of their fall quarter as CNI fellows coordinating an elective course titled "Women, the Environment, and Physician Activism: Encouraging Activism through Education." This course consisted of ten 1-h-long lecture sessions given by a wide range of professionals on topics that included inequality, prenatal health, and sustainable food practices. Dr. Robert Gould, who was the faculty adviser for this course, has been a key player in pushing these topics into the health field and education. "He's essentially working on how health professionals should be involved in these issues. Health professionals are respected by society, so how can we use that power for good? He has really helped us organize the class," explained Carolyn.

Dr. Gould leads a session called Generations at Risk: Introduction to Environmental Health and Health Professional Activism. The course was offered to first- and second-year medical students as an OB-GYN elective. "I initially got into this because I am very interested in going into OB-GYN and so the connection between woman's health and potential reproductive justice and environmental impacts was very interesting," said Gabriela. She continued, "Pregnant woman are way more likely to be concerned about environmental issues than just the normal, everyday person."

Because of the heightened levels of concern, topics regarding health and the environment are first being integrated into OB-GYN and reproductive health curriculum. "It's kind of like a pilot run to integrating it more broadly," Carolyn described. Through courses like these, the hope is to raise interest and create demand to have these types of conversations in other classes as well. The CNI fellows explained that students are asked to take pre- and post-course surveys, which "assesses comfort and motivation to talk about the issues, and whether or not students think they are important."

However, a course that highlights the effects of climate change and environmental dangers runs the risk of being more depressing than motivational. "We want to provide concrete ways to approach a very ominous and kind of scary topic, using the power that comes with being a health professional and the voice it gives you," said Gabriela. "We want to leave students with action items and use case-based learning where the outcomes were positive," added Carolyn.

To achieve this, speakers were asked to talk about specific action items that students could do or provide them with helpful resources. For example, Dr. Mark Miller, UCSF Assistant Clinical Professor in the Departments of Pediatrics and Internal Medicine (Division of Occupational and Environmental Health),

Co-Director of the Western States Pediatric Environmental Health Specialty Unit (PEHSU) at UCSF, and the director of the Children's Environmental Health Center at the California Environmental Protection Agency, presented about pediatric environmental health and made available an online toolkit that he worked to create with the PEHSU staff. The toolkit assists clinicians in incorporating preventive environmental health messages into routine pediatric care. He is also the primary author of *The Story of Health*, a multimedia E-book that promotes the environment and health through storytelling.

In addition to the ten speaker sessions, the final course session was a showing of the documentary *Merchants of Doubt*, a film which identifies causes of misinformation and public confusion in regard to climate change. It highlights parallels between groups involved in the tobacco industry and those pushing for oil and denying climate change. "The documentary does a good job at explaining that science doesn't always do very well at communicating to the public," said Gabriela. Carolyn added, "The people who are perpetuating climate change are really good at this stuff, and although we maybe have the science we aren't very good at communicating. This is kind of the issue of our time and to not train health professionals in it seems crazy to me. You can't just rely on Leonardo DiCaprio for everything."

While it's not impossible to avoid the doom and gloom, empowerment enables a sense of hope and helps to remove the barriers of fear and denial that come with negative environmental concerns and the health factors that accompany them. Everyone needs to be empowered to make change or change won't happen. Through education and providing response tools to tough issues, students can be empowered to speak and act on these topics as health professionals and in their everyday lives.

Case Studies for Medical Facilities in the United States

The two green healthcare case studies that follow are from the Healthier Hospitals organization http://healthierhospitals.org/get-inspired/case-studies.

University of Massachusetts Medical School Case Study: Central Plant Energy Efficiency

Much of the following case study information was taken (with permission) from the following article (http://www.healthierhospitals.org/sites/default/files/IMCE/umass_medical_school.pdf).

Summary

The CHP (combined heat and power) project was undertaken to better support a growing campus. To make this project financially possible, UMass Medical School combined current construction efforts to include a new combined heat and power (CHP) system and received financial incentives from the local utility company. The project saved annually 58,000 MWh in electricity and $6.2 million.

The Problem

In 2010, UMass Medical School (UMMS), including UMass Medical School and UMass Memorial Medical Center, had an existing and sophisticated central plant that provided steam, chilled water and power to the entire campus. However, they needed to support its growing campus, specifically the energy needs of the planned Albert Sherman Center, a $300 million 512,000 ft^2 biomedical research and clinical education facility, scheduled to open in January 2013.

The Strategy Selected

UMMS's long-term campus and utility master plan envisioned the continued expansion of combined heat and power on their campus. With the inception of the Sherman Center, UMMS began programming this CHP expansion project to include looped chilled water and steam distribution systems as well as the equipment diversity such as steam drive and gas turbine drive generators. In order to accomplish this financially, they included the upgrade as part of a larger $450 million campus capital campaign and integrated it into plans for new construction. This allowed UMMS to capitalize on low-cost financing and also ensured a fully integrated implementation approach. UMMS installed a new 7.8 MW gas turbine with 60,000 pph heat recovery steam generator to support the campus's existing and future electrical, steam and chilled water loads. The system was also designed with enough capacity to support future expected construction at the campus.

Implementation Process

Leadership from the school and hospital supported the strategy and project on all accounts – infrastructure resiliency and redundancy, greenhouse gas reductions, energy efficiencies, and life cycle cost reductions – making the project possible.

UMMS's project team completed the construction and installed a Taurus 70 Solar Turbine – a mechanical drive package, which can be combined with one or more centrifugal gas compressors to form a complete compressor set. Designed specifically for industrial service, Taurus 70 packages are compact, lightweight units requiring minimal floor space for installation. The project also included an electric drive 4000 ton chiller, two new cooling towers, and associated electrical switchgear. The CHP system was installed as part of a $48 million overall expansion to the campus's existing central plant, which was offset by $7 million from a National Grid incentive.

Benefits

Increased savings, power, and energy efficiency were realized through this project including:

- 58,000 MWh in annual electricity savings
- $6.2 million in annual savings
- Less than 3-year payback period
- Increased power production and chiller capacity for the 500,000 ft^2 Sherman Center
- Designed to support future construction at hospital
- Backup for the hospital's existing central plant

Challenges and Lessons Learned

The central plant at the UMMS campus is a complex and sophisticated system of multiple technologies. Effectively integrating the new gas turbine into this system and optimizing system-wide performance was a complicated engineering task. Additionally, given the scale of the system, interconnecting the CHP unit into the local electrical grid was a challenge. UMMS staff had significant experience working with National Grid on previous projects and were able to work closely with utility representatives to meet all interconnection requirement.

School staff noted that incorporating the CHP system into the early planning phase of the hospital's long-term master plan was a major reason for the success of the project. This long-term view of CHP project development allowed the hospital to take a careful and considered approach to CHP system planning. Staff also noted that a close working relationship with utility staff was a critical success factor for the project.

Pointers to Other Chapters for Additional Details on Case Study Lessons Learned

This case study emphasized the need for collaboration, central plant efficiencies, etc. Here are pointers to the chapters that give additional detailed information on those topics:

- Collaboration. Chapter 7
- Green IT. Chapters 5 and 7
- Network Design and Data Center Efficiency. Chapters 3 and 5
- Building Automation Systems. Chapters 3 and 5

Kaiser Permanente Case Study: Purchasing Environmentally Responsible Computers

Much of the following case study information was taken (with permission) from the following article (http://healthierhospitals.org/get-inspired/case-studies/kaiser-permanente-electronic-products-environmental-assessment-tool-epeat-).
Electronic Products Environmental Assessment Tool (EPEAT)

Benefits

- Environmental and Human Health Impact: Vast reduction in use of toxic materials (lead, cadmium, and mercury) and energy, increased use of recycled resins, recycled content packaging, and reusable packaging.
- Business Impact: Up-front purchase of computer systems is cost neutral with energy cost savings of $4 million per year.

The Problem

Computers have enabled Kaiser Permanente to provide our members, patients, and physicians with real-time access to electronic medical information which has expedited and simplified delivery of care. But the manufacture, use, and disposal of computers and their electronic accessories have a global adverse impact on human and environmental health. Recognizing the benefits and problems with electronics, Kaiser Permanente was looking for a way to identify more environmentally responsible computer systems. The contract with a new computer system supplier in early

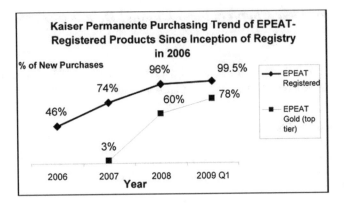

Fig. 10.1 Kaiser Permanente purchasing trend of EPEAT-registered products

2006 included language specifying a strong and definite preference for energy efficient and environmentally responsible computers. Figure 10.1 shows the purchasing trend, starting in 2006.

The Strategy Selected

- To procure CPUs, monitors, and laptops that are manufactured with least toxic materials, are designed for easy recycling, use minimal energy for operation, and are packaged with minimal material
- To reduce KP's energy consumption and costs

Implementation Process

The Team:
- IT procurement sourcing team
- Pradeep Saxena, sourcing director, IT

- In 2006, the EPEAT criteria and registry system that enabled an easy comparison of the environmental performance of computer systems were launched.
- Prior to EPEAT, Kaiser Permanente utilized the Energy Star® rating system as a minimum requirement for new computer systems. However, procurement and supply's IT sourcing team quickly adopted EPEAT due to its comprehensive focus on environmental issues, which includes energy performance.

- Expectations were set with the incumbent supplier that as new products meeting higher registry tiers became available, Kaiser Permanente would experience swift adoption of those models in concert with our purchasing standards.

Next Steps: Continue standardization on purchasing solely EPEAT gold products when product refresh is required. (EPEAT gold products meet all required and 75% of optional criteria.) Participate in working groups and support the expansion of EPEAT to new product areas including print imaging devices and televisions.

Challenges and Lessons Learned

- The use of a reputable and verifiable third-party certification tool can simplify the purchasing of environmentally preferable products.
- A certification system, like EPEAT, that compares supplier performance in a transparent and balanced fashion can help prevent price markups and "green washing" (disingenuous promotion or advertising of product environmental attributes).

Pointers to Other Chapters for Additional Details on Case Study Lessons Learned

This case study emphasized the need for collaboration, green IT, etc. Here are pointers to the chapters that give additional detailed information on those topics:

- Collaboration. Chapter 7
- Green IT. Chapters 5 and 7
- Network Design and Data Center Efficiency. Chapters 3 and 5
- Building Automation Systems. Chapters 3 and 5

Six Sigma Case Studies

The Six Sigma process for making healthcare green was described in Chap. 2 of this book with this reference (Franchetti, 2013). Several important Six Sigma healthcare case studies are given in the following article (https://www.isixsigma.com/implementation/case-studies/).

A synopsis for each of four of those healthcare case studies are listed below.

A Successful Pain Management Initiative at LDS Hospital

Despite already having adopted many best practices, LDS Hospital in Salt Lake City discovered a problem in the rapid treatment of pain in the emergency department. The staff immediately began a pain management initiative to remedy the problem.

Achieving and Sustaining Improvement in Cardiac Medication

A Six Sigma cardiac medication project was successful in coordinating high-quality care, incorporating advanced technologies, and adopting best practice standards.

Addressing the Issues at Doctors Hospital of Augusta.

Implementing Six Sigma has allowed Doctors Hospital of Augusta to begin winning the battles with the familiar healthcare issues related to quality, resource constraints, and ensuring optimal, accessible services for the community.

Applying Six Sigma to Improve Diagnostic Imaging

Six Sigma can help the traditional workflow of diagnostic imaging keep pace with the science delivering better equipment and information systems (Arthur, 2011).

Attention Six Sigma, You're Wanted in Surgery

With the demand for surgical services growing, especially with ambulatory and cosmetic procedures, hospitals continue learning that Six Sigma is a useful tool.

Chapter Summary and Conclusions

The case studies described in this chapter included "lessons learned" for medical colleges and large and small hospitals in the United States. As discussed at the beginning of this chapter, universities offer a rich environment for green healthcare innovations and best practices. Not only do they have the administrative systems common to any large healthcare facility, but with their university research departments, they have many groups of researchers in varied fields who can help improve the university healthcare energy efficiency and carbon footprint.

Of course, large hospitals (including university hospitals) typically have many more resources for making the hospital green. However, the significant value in looking at green healthcare case studies, for both large and small healthcare facilities, should be to take the opportunity to take advantage of the case study lessons learned. Chapter 11 presents case studies for healthcare facilities worldwide.

Additional Reading

Muduli, K., & Barve, A. (2012, August). Barriers to green practices in health care waste sector: An Indian perspective, *International Journal of Environmental Science and Development, 3*(4), 393–399.

Chapter 11
Worldwide Case Studies for Green Healthcare

What we have to learn to do we learn by doing.

—Aristotle, Ethics

Be the change you want to see in the world.

—Mahatma Gandhi (India)

The challenges in implementing green healthcare and green hospitals can vary considerably based on the part of the world where the healthcare facilities are located. Challenges include government regulations, the cost of electricity, and the social/political environment relating to the environment and energy efficiency. The first few chapters discussed the different aspects of green healthcare. The case studies in this chapter include green healthcare for countries around the world, including countries in Europe, Asia, Africa, and North and South America. Case studies for hospitals in the United States were discussed in Chap. 10. Although the social and political environments of the particular country do have an impact on implementing green healthcare, the technical challenges remain the same. The significant value in looking at green healthcare case studies should be to take advantage of lessons learned.

Global Green and Healthy Hospitals (GGHH)

Much of the following case study information was taken (with permission) from the following article.

https://noharm.org/sites/default/files/archivednewsletters/HCWHGlobal/gghh_newsletter/GGHH_DEC_2013.html.

Global Green and Healthy Hospitals has 826 members in 48 countries on 6 continents who represent the interests of over 27,800 hospitals and health centers. Their members are using innovation, ingenuity, and investment to transform the health sector and foster a healthy future for people and the planet. The network is based on the members' commitment to implement a comprehensive environmental health

© Springer International Publishing AG, part of Springer Nature 2018
N. S. Godbole, J. P. Lamb, *Making Healthcare Green*,
https://doi.org/10.1007/978-3-319-79069-5_11

framework for hospitals and health systems called the Global Green and Healthy Hospitals Agenda.

Founding members of the network, which HCWH coordinates, include the Sustainable Development Unit, NHS England and Public Health England, Thailand's Department of Health, the Healthier Hospitals Initiative in the United States, the Indonesia Hospital Association, the Australian Healthcare and Hospitals Association, the International Health Promoting Hospitals Network, FHI360, and dozens of individual hospitals from Australia, Brazil, China, Chile, Colombia, France, Germany, India, Nepal, South Korea, Taiwan, South Africa, and more.

Argentina

Waste: Argentina | Good Waste Management Practices in Healthcare Establishments, Dr. Roque Sáenz Peña Hospital in Rosario.

United Kingdom

Waste: United Kingdom | Reducing the Carbon Footprint of Medicinal Waste Disposal, Newcastle upon Tyne Hospitals NHS Foundation Trust.

Philippines

Water: Philippines | Hospital Engineers Discharge Cleaner Water, Philippine Heart Center.

Brazil

Purchasing: Brazil | Reducing Suture Stitches Packaging, São Paulo's Hospital Israelita Albert Einstein.

The case studies in the following pages from South Africa and Argentina were taken from the GGHH web site. These case studies are followed by a case study from the United Kingdom (documented on the Carbon Trust web site). The UK case study is followed by three green hospital case studies in India based on numerous visits to the hospitals by author Nina Godbole.

South Africa Case Study: Hospital Solar and Wind Energy System

Much of the following case study information was taken (with permission) from the following article.

http://greenhospitals.net/wp-content/uploads/2015/06/Khayelitsha-District-Hospital-South-Africa.pdf.

The Khayelitsha District Hospital in South Africa (Western Cape area) makes use of a solar and wind grid-connection power plant that is connected to the electrical system and was commissioned on August 30, 2011. The aim is to significantly reduce energy consumption and carbon emissions.

Savings

From August 30, 2011, to January 14, 2014, the solar and wind plants generated 138 mWh of electricity. This equates to 137 tons of carbon emissions having been avoided. This corresponds to the CO_2 emissions of a car traveling a distance of over 1 million km (650,000 miles). Furthermore, the hospital staff has been brought on board to switch off all unnecessary electrical appliances and air conditioners. The saving of the solar and wind plant, together with the awareness of switching of electrical appliances, has resulted in a financial saving of about R200,000.00 per month. The Khayelitsha District Hospital was the first hospital in the provincial government Western Cape that is equipped with a green initiative such as solar photovoltaic system and a wind energy system.

Sustainability Strategy Implemented

South Africa experiences frequent load shedding due to major maintenance problems at power plants. By installing the 25 kWp solar system consisting of 108 solar panels on the roof of the hospital, and 2 wind turbines each with a capacity of 2 kW wind energy, the hospital managed to reduce electricity use.

Implementation Process

The photovoltaic and wind system has been installed by Power Solutions (www.powersolutions.co.za). Their senior engineers are registered as professional engineers with the Engineering Council of South Africa. They are also members of South African Institute of Electrical Engineers, South African Photovoltaic Industry Association, Green Building Council of South Africa, and the South African Alternative Energy Association. This system generates DC voltage up to 900 V and AC voltages of 400 V.

The 108 solar panels cover an area of 190 m². They generate electricity during sunshine hours and feed into the electricity grid. At night it goes into standby mode. This system does not export power into the municipal grid; therefore Khayelitsha hospital is a net electricity importer. The power generated is about 25 kW at noon on a clear day and that is approximately 40 MWh per year. The warranty on this system is for a life expectancy of 25–30 years during which time it will provide at least 80% of the current output.

The wind energy system consists of two vertical axis wind turbines (VAWT) with a safety brake switch, windy protection box, diversion load, and WindyBoy grid inverter. The WindyBoy inverter converts the power generated by the VAWT to make it compatible with the electricity grid. The voltage first needs to be rectified by the windy protection box. The safety brake is used to stop the system when the wind blows too hard (in excess of 40 m/s). This system is monitored by the Sunny WebBox, which logs all the data of the system and sends it via the local network to an online portal. In this case Power Solutions monitors all the data.

Tracking Progress

The system was also installed with a Sunny WebBox. This WebBox is a system monitoring, remote diagnostics, data storage, and visualization and serves as the communication center for the solar power station. It gathers information from the system side and allows the hospital to be continuously informed of the status at any given time. This system has a GSM modem and therefore doesn't need DSL or telephone connection to communicate from the remote location. It further has a large-scale display unit in the main foyer of the hospital for everyone to see. This displays the total amount of energy generated to date, the current power output and total amount of carbon emissions avoided.

Challenges and Lessons Learned

Having a state-of-the-art hospital comes with challenges, especially involving energy. Basically everything that is installed in the hospital is designed to ensure that patients and staff have enough fresh air that circulates in the areas; however, in order to maintain the desired air changes per minute, an air-conditioning system needs to be installed. Running this air-conditioning system requires electricity. In other words, it is all very nice equipment, but in plain terms it would have been just as effective to rely on the wind and open windows for the sufficient air changes instead of an electrical-driven air-conditioning system that requires full-time maintenance and electricity that is high in cost. The same applies to the fancy electrical turnstile entrances. It takes unnecessary energy, and therefore it is switched off to save on electricity, and the public uses the manual entrances that remain open 24/7. Architects sometimes concentrate so much on the design and how the building presents itself to the public that they tend to forget that simpler systems would also have worked, will require less maintenance, and will be more economical, especially with government building where supply chain processes have unnecessary red tape.

Pointers to Other Chapters for Additional Details on Case Study Lessons Learned

This case study emphasized the need for collaboration, green IT, etc. Here are pointers to the chapters that give additional detailed information on those topics.

- Collaboration. Chapter 7
- Green IT. Chapters 5 and 7
- Network Design and Data-Center Efficiency. Chapters 3 and 5
- Building Automation Systems. Chapters 3 and 5

Argentina Case Study: Best Practices in Waste Management

Much of the following case study information was taken (with permission) from the following article.

https://noharm.org/sites/default/files/archivednewsletters/HCWHGlobal/gghh_newsletter/GGHH_DEC_2013.html.

The Objective of This Global Agenda of Green and Healthy Hospitals Project

To reduce, to treat, and to dispose the waste in a safe manner the waste (generated) in health establishments.

The date of execution (of the waste management project): September 2012.

Benefits

- Saving of approximately US $25,000 per annum, considering that the (waste) separation best practices keep assuring that 21,000 kg of waste would not need any special treatment.
- Reduction in bio-contaminating (infectious) waste in the range of 24,610 kg/annum to 44,421 kg/annum due to the implementation of norms for the classification of waste within an institution.
- Reduction in work-related accidents to due sharp pointed objects.
- Creation of a horizontal and participative management: created a discussion space regarding this theme in spite of the difficulties faced. It involved the participation of all the actors and related parties including the operating management level.

- Awareness about the participation in problems of all the sections of the hospital.

The Problem

The waste generated in the hospital becomes a serious problem until its final disposal if it is not classified and not treated adequately. To address this, within the Public Health Ministry of Rosario and the Ministry of Public Services, a project of continuous work and the Project of Rosario Waste were initiated, through which the Internal Management Committee for the waste was established. The committee counted on the support from the director general of the hospital.

Objectives

- To reduce environmental pollution
- To boost biosafety
- To reduce costs
- To encourage training programs and continuing education

Methodology

1. *The launch phase*:

    (a) Through an institutional and organizational analysis, a diagnosis of the situation was completed.
    (b) Medical services were launched.
    (c) A study was made about the generation of bio-contaminating (infectious) waste.
    (d) The composition of bio-contaminating (infectious) waste was studied.
    (e) A planning workshop was undertaken.
    (f) Norms were drafted and posters put up (see Fig. 11.1).

2. *The implementation phase*:

    (a) Workshop for the classification hospital waste
    (b) Development of general norms for the classification and refurbishment of hospital waste
    (c) Validation of the norms by the director general of the hospital and vigilance of the norms starting from October 2004

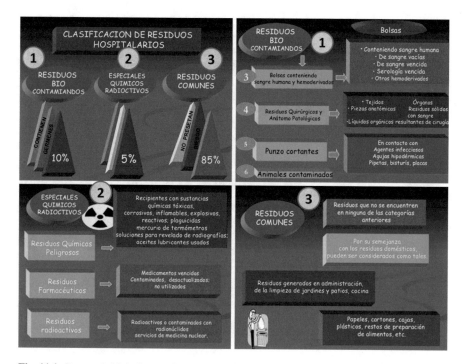

Fig. 11.1 Posters (with indicators for the separation) located in passages and wards

Implementation Process

The committee dictates the general guidelines according to the concerned legislation. Each service tailors the guidelines for best functionality. The project was made available for all the services/sections of the hospital. The norms for (waste) classification (see Fig. 11.1) followed the workshop in which a consensus was achieved on the criteria for the (waste) classification and its application later. A training schedule was prepared for the general training for all the personnel. In addition, on-site training was implemented. At present, too, annual training is implemented along with records of the training maintained.

They carried out "orientation talks for the newly registered personnel"; these talks consist of:

- Concepts of biosafety
- Accidents at the workplace
- Launching of vaccines

Given that it is a teaching hospital (i.e., a health facility attached to a university), there are residents and students who are in their final years of study, recently joined doctors who are doing their professional practice in medicine, students of infirmary at the National University of Rosario, and interns and scholars (i.e., those on schol-

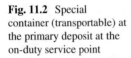

Fig. 11.2 Special
container (transportable) at
the primary deposit at the
on-duty service point

arships) of any (medical) service. Training is offered to all these personnel and to all
the new personnel who join the hospital, regardless of what their functional department may be.

Modification of the Facilities

- Placement of containers at the points of segregation of urban solid waste (the
  common one) and bio-contaminating waste (infectious)
- Provisions of containers holding sand for discarding cigarettes
- Placement of special containers of bigger size which are transportable, for the
  intermediate storage at the site (see Fig. 11.2)

- Cleaning/refurbishment of the biohazard waste containers after being emptied
- Adaptation of a new receptacle for the hazardous chemical waste
- Placement of containers for the recyclable waste (see Fig. 11.3)
- Creating routes for the gathering of each category of waste
- Utilization of carts/trolleys for the collection and internal transportation of dif-
  ferent types of waste (see Fig. 11.3)
- Establishing procedures
- Placement and utilization of signposting/signage/markers

Fig. 11.3 Containers for hospital waste collection

Challenges and Lessons Learned

Factors that presented difficulties in this process:

- Initial lack of awareness about the theme
- Resistance to change
- Lack of commitment on the part of certain members of the hospital staff
- Lack of continuity in the provision of consumables

Factors Favorable to this Process

- Commitment from the director of the hospital
- Support from the infirmary department
- Commitment from the hospital personnel
- Studies about the generation and composition of the bio-contaminators (infectious)

From the very beginning, the management is horizontal and participative, and that permits to a great extent a sense of belonging (to the project) in the staff. There is room for offering suggestions that get jointly evaluated for their applicability. Working on national and international projects is also beneficial, as well as the hold-

Fig. 11.4 Posters and containers differentiated for (various types of) waste: *red* for bio-contaminating waste, *black* for urban solid waste, and *white* for saline bags to be recycled

ing of periodic meetings as also the permanency of members of the committee. While there have been times of great enthusiasm, there also have been those of low spirits; but they were overcome by consolidating the working teams. The years of uninterrupted work have made the hospital recognized as a referral site for the topic of waste.

Next Steps

Work is in progress for:

1. The improvement of some of the intermediate deposits of bio-contaminating (infectious) waste (see Fig. 11.4)

2. Reconditioning of the containers for hazardous chemical substances
3. Washing and maintenance of the containers utilized as intermediate storage
4. Revision of procedures
5. Updating the classification norms

Demographic Information

Dr. Roque Sáenz Peña Hospital is located in the south district of Rosario (province of Santa Fe, Argentina). It is a 98-bed second-level facility, responsible for 2 of the city's 6 districts and 20 primary care centers.

Pointers to Other Chapters for Additional Details on Case Study Lessons Learned

This case study emphasized the need for collaboration, etc. Here are pointers to the chapters that give additional detailed information on those topics.

- Collaboration. Chapter 7

United Kingdom Case Study: University of Reading

Much of the following case study information was taken (with permission from the Carbon Trust) from the following article:

> https://www.carbontrust.com/our-clients/u/university-of-reading/.

It should be noted that this article was originally published in May, 2015, and progress had been made since that date. Also, the Carbon Trust has developed carbon management plans for a number of UK hospitals using a similar approach.

Identifying Carbon Reduction Opportunities

In 2010, the University of Reading designed and implemented a Carbon Management Program with the assistance of the Carbon Trust. Their motivation to reduce emissions through this program was partly triggered by the announcement of the Higher Education Funding Council for England's goal of a sector-wide 43% reduction in carbon emissions by 2020 and partly for financial reasons, in recognition of spiraling energy costs. To kick off the program, a Carbon Management Plan was drawn up, identifying carbon reduction opportunities and setting an overall emissions reduction target of 45% by 2020, with an interim target of 35% by July 2016.

The development of this Carbon Management Plan was essential to making the business case for carbon reduction, identifying that a £3.5 m investment in energy efficiency up to July 2016 could yield £18.5 m in cumulative savings. It also ensured that senior management bought in to the program, including endorsement of the program and the carbon reduction targets by the vice chancellor. A Carbon Management Board chaired by the then deputy vice chancellor was established and is supported at implementation level by the University's Carbon Management Team.

This governance structure has been a key facilitator of progress throughout the program. Dan Fernbank, Energy Manager at the University of Reading, said, "the governance structure established in the Carbon Management Plan was integral to achieving 23% carbon reduction by July 2014, through ongoing senior management support, a clear decision making process and raising the profile of the program." The business case for investment has been regularly reviewed, updated, and reported against to track the progress of the program. To date, £2.35 m has been invested and has provided a £8.5 m saving for the University and its partners.

Collaborative Implementation

The Collaborative Implementation Program ran from 2012 to 2013, providing workshops and training and bespoke support to implement high-level projects identified in the Carbon Management Plan. This year-long program helped to explore implementation barriers and provide solutions to overcome them, to help participants learn to solve problems by themselves when implementing carbon reduction projects. The presence of industry experts delivering this training and support gave participants an opportunity to ask direct technical questions, as well as providing them with a better understanding of the considerations for different technologies.

Dan Fernbank said, "the Collaborative Implementation Program really helped to turn high-level ideas into practical solutions, meaning we've been able to deliver year-on-year cost and carbon savings."

Broadening the Scope

The University found that the key to progression was to keep the scope of projects broad, rather than concentrating on only a few common technologies. This diversity meant that carbon was reduced from all angles, including efficiency improvements, BMS controls, and technology upgrades. Example projects include:

Cost and CO_2 Reduction

- Insulation program – plant room pipe lagging, roof insulation, and draught proofing saving 1100 ton CO_2 annually
- IT server upgrades – saving 1100 tons of CO_2 annually
- Lighting upgrades – efficient lighting with intelligent sensors, saving 775 tons of CO_2 annually
- BMS/controls expansion and upgrades – saving 550 ton CO_2 annually
- Fume cupboard ventilation upgrades – saving 500 ton CO_2 annually
- Heating plant/control upgrades – saving 600 ton CO_2 annually
- Ventilation and air-conditioning upgrades – saving 400 ton CO_2 annually

In addition to these specific interventions, there has been a major program to replace aging halls of residence with new facilities, including the refurbishment of the London Road campus and the closure of the Bulmershe campus and two off-campus halls of residence. This led to savings of approximately 4000 ton CO_2 annually while simultaneously increasing the total halls of residence bedroom numbers.

Overcoming Barriers to Success

Prioritizing projects across a large campus has been a key challenge throughout the program. The University lacked comprehensive building-level energy metering at the beginning of the program, so in 2012 a metering strategy was set out for the estate. Since then, £350,000 has been invested in building-level electrical, heat, and water meters in all principal buildings on the estate, and an estate-wide energy review was carried out. These initiatives allowed a prioritized program to be drawn up for both surveying and implementation work and are also enabling focused behavior change campaigns by engaging building occupants with the energy use of their buildings.

The business and estate is also constantly evolving, so it has been crucial to understand the potential impacts and benefits of projects and initiatives beyond the direct control of those implementing the Carbon Management Plan. This has led to the development of a "Carbon Trajectory" tool to link in with the University's existing Carbon Plan monitoring to project emissions each year to 2020.

The Future

As the 35% carbon reduction target draws closer, the University of Reading is confident that it is on track. University staff are currently developing more detailed plans for progress to the 45% 2020 target. Major projects in the pipeline include the full implementation and optimization of a new district heating network, further fume cupboard efficiency work, comprehensive science building energy audits, and a large-scale solar PV scheme.

Pointers to Other Chapters for Additional Details on Case Study Lessons Learned

This case study emphasized the need for collaboration, etc. Here are pointers to the chapters that give additional detailed information on those topics.

- Collaboration. Chapter 7
- Building Automation Systems. Chapters 3 and 5

Case Studies in India

A Green Hospital in India: IGBC Case Study

The "going green" concept is on rise in India, the country which has become one of the top destinations as far as medical tourism is concerned. In the recent years, India has witnessed the green building movement gaining tremendous impetus. This reflects in the fact that the market for LEED-rated green buildings in India rose to $5 billion in 2012, and the total market for green building materials and products is estimated to be over ten times (around $45–50 billion) the size of the LEED-rated green building market in India.

Indian Green Building Council (IGBC)

The Indian Green Building Council (IGBC) has been active in the recent years as well as other Indian green-movement agencies, for example, GRIHA (Green Rating for Integrated Habitat Assessment). It is a green building rating system developed indigenously, keeping in mind the Indian construction environment. It was developed by The Energy and Resources Institute (TERI) and has been adopted by the Ministry of New and Renewable Energy (MNRE) as the national green building rating system for India. GRIHA is aligned with various national building codes and guidelines like the National Building Code, Energy Conservation Building Code, Ministry of Environment and Forests clearance, Central Pollution Control Board guidelines, etc. The Energy and Resources Institute (TERI) has taken considerable efforts as far as the green habitat is concerned. As mentioned, TERI has developed a scheme for green evaluation (known as TERI-Green Rating for Integrated Habitat Assessment).

In the last few years, green building construction has made significant inroads in India. Today it has emerged as one of the world's top scenes for green buildings (ranging from residential complexes, exhibition centers, hospitals, and educational institutions to laboratories, IT parks, airports, government buildings, corporate offices, etc.) and implemented a number of rating systems (LEED-India, GRIHA, etc.). This has opened up a wide range of opportunities in areas such as architecture and engineering design, construction, equipment manufacture, and building materials. In India, a large variety of buildings are registered for LEED Green Building Rating, including IT Parks, offices, banks, hospitals, airports, convention centers, educational institutions, hotels, and residential buildings. The Green Building Movement in the country is pioneered by the Indian Green Building Council (part of CII), and currently, India has 1239 registered buildings and 187 certified buildings with 855.96 million ft^2 Green building footprint.

As mentioned in Chaps. 3 and 5, traditionally, in most hospitals the focus of green has been in solid waste management rather than the management of the electronic waste (eWaste). The thought of including eWaste in a hospital's green efforts is a relatively new phenomenon. While this may be true, India has a number of hospitals (apart from the hospital whose case study is presented here) that have registered themselves with the IGBC (Indian Green Building Council) for the "green" rating.

The Super Gem Hospital

With this green background, we present a small case study of a sustainable hospital in India. It is the Super Gem hospital whose real name has been masked for privacy reasons. The Super Gem hospital is located in one of the suburbs of Mumbai, in the Maharashtra state of India. The Super Gem Hospital is a 227,000 ft², 150-bed facility. It is housed in a building that has two basements and five floors. The hospital has obtained the highest rating possible (i.e., the platinum status) that is available under the LEED certification scheme. The Super Gem Hospital is the world's second LEED platinum-certified hospital and Asia's first. This hospital has been awarded 54 credit points and is the only LEED-certified hospital in Mumbai, India.

The management of Super Gem Hospital understands that patient recovery is much faster when the hospital has the connectivity with its external environment; the daylight is better and the quality of the indoor air is better. In this multispecialty hospital, a range of energy-efficiency measures have been implemented with the main objectives of reducing the overall energy consumption and decreasing GHG emissions. Of the construction material used for Super Gem Hospital, more than 40% was the recycled material. The foundation of the overhead tank was built with reused scrap material, and frames were made by using salvaged wood; this helped to conserve trees.

The hospital design has incorporated the use of natural lighting in the patient areas in order to reduce the consumption of electrical energy. The large windows and open skylight not only keep the patient areas cool and ventilated but also connects the patients with their external environment. In the hospital building complex, recycled water is used for flushing toilets, cooling of the air-conditioning towers, diesel generator sets, and also for the horticulture. This approach has helped them save 40% of the water.

The hospital has insulated walls with very low "U" value. Its energy use intensity is 53 kBTU per ft² per year. There is an optimal window-to-wall ratio and uses shaded windows. High-performance glass has been used for the windows of Super Gem Hospital. The hospital's artificial lighting constitutes 20% of the overall energy load. This has resulted in 50% savings. The use of LED and low LPD (lighting power density) results in substantial reduction in internal heat gain. The hospital has an efficient system for handling the waste generated. One hundred percent of the waste produced is recycled or reused or given away for use by the local community. As compared to other building constructions, the hospital circulates 30% more pure air.

The time switches used in the hospital are programmable basis of the timing of sunrise and sunset. These switches function without the use of light sensors. This not only saves the amount of electrical energy but also increases the life span of lamps, thereby further reducing the solid waste generated in the hospital. By virtue of being "green," the building of Super Gem Hospital consumes 0.66 W/ft^2 as compared to a normal construction that would consume 2.0–2.5 W/ft^2. The green roof used for the hospital building provides insulation from the outdoor environment. Solar panels are installed on the top of the hospital building to harvest the solar energy derived from the sun. This solar energy is used in Super Gem Hospital to heat water and to maintain the required humidity level in the operating theaters.

The HVAC (heating, ventilation, and air conditioning) design of the Super Gem Hospital includes chilled water plants which consist of energy-efficient screw chillers. These chillers are driven by VFDs (variable frequency drives). The building design of Super Gem Hospital is based on integrated building design system. All parameters of the HVAC are controlled by the BAS (building automation system). The BAS also controls electrical services utilized in the hospital, its elevators, and fire protection system. The hospital's direct digital control (DDC) system interfaces with sensors, actuators, and environmental control systems to carry out various functions of energy management, alarm detection, time/event/holiday/temporary scheduling, communication interface/control, and building maintenance and report generation. The electrical system of the Super Gem Hospital uses the latest technologies and fundamental principles of energy conservation and safety to provide protection against thermal effects, electric shocks, over current, and fault current as well as protection against over voltage. The use of light sensors is made in the hospital building to switch off lighting whenever space is not occupied or whenever natural lighting is available. The use of ELCB (earth leakage circuit breakers) is meant to safeguard against electrocution to accidental contact with a live current.

Next Steps

The hospital is now in the process of considering the use of cloud technologies and thin servers to see if further energy conservation is possible in the ICT systems that they use.

Pointers to Other Chapters for Additional Details on Case Study Lessons Learned

This case study emphasized the need for collaboration, green IT, etc. Here are pointers to the chapters that give additional detailed information on those topics.

- Collaboration in general. Chapter 7
- Collaboration with groups such as TERI. Chapter 7
- Green IT. Chapters 5 and 7
- LED for Lighting. Chapters 1 and 10

- LEED. Chapter 9
- Telemedicine. Chapter 3
- Network Design and Data-Center Efficiency. Chapters 3 and 5
- Use of EMR and HER. Chapters 2 and 3
- Building Automation Systems. Chapters 3, 5, and 7
- Variable Frequency Drive for heating and cooling systems. Chapters 2 and 4

A Green Hospital in India: HCwBQ Case Study

Clad in Green: The Case of an Indian Hospital Embracing the Green Strategy

The Background: HCwBQ (Healthcare with Best Quality)

Way back in 2005, the proactive management of HCwBQ hospital (Healthcare with Best Quality – referred to as *HCwBQ* throughout the case study) realized that the modern healthcare sector not only contributes to GHG generation and its share in the carbon foot print but is also itself affected by it. The young and dynamic CEO, at that point of time, was just back home after completing a program in sustainability management at an overseas university.

The case study, presented here, was undertaken to understand (1) the "green vision" of the hospital's top management team and (2) how they put it in practice to embrace green in HCwBQ. This case study presents the meaning of "green hospital" as understood at *HCwBQ*.

In the first interview with the young CEO, when we asked him to explain what a "green hospital" meant to him, his response was: "A Green Hospital is one which enhances not only the well-being of its patients and the best healthcare outcomes (diagnostic and curative processes) but also utilizes natural resources in an energy-efficient and environmentally friendly manner with the main objective to achieve reduction in the consumption of energy, reduction of costs related to energy (among other costs) and reduction of CO_2 emission levels." He further added that from an implementation perspective, HCwBQ has looked into the *Standard for Green and Clean Hospital*, the standard prepared by AHPI Technical Committee. AHPI is Association of Healthcare Providers (India). The standard provides a number of guidance notes.

HCwBQ is a multispecialty hospital with 300+ beds to provide high-quality care. Like most other hospitals, it is a resource-intensive establishment and consumes vast amounts of electricity, water, and food. Given its expansion plans, it would also consume construction materials. During the investigative study of *HCwBQ*, we found that by employing a few simple, smart, and sustainable measures, *HCwBQ* has been able to reduce their environmental footprint to a considerable extent.

The critical part in the formation of HCwBQ's green policy was the keenness of its management to understand the components that impact or contribute to hospi-

Fig. 11.5 *HCwBQ* road
map

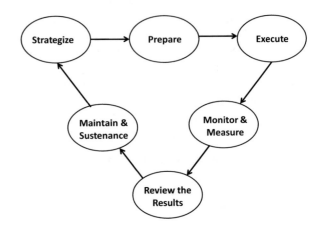

tal's carbon footprint. The top management team, during one of the interviews, identified these to be (1) energy consumption, (2) building design, (3) waste management, (4) water conservation, (5) transportation system, and (6) providing healthy food.

Another important consideration for the management team of *HCwBQ* was to also look into the IT systems of the hospital from the perspective of greening it, especially the servers and their dedicated data center. Identification of these elements helped in the formation of their green policy-based strategy. One more important component of the go-green strategy in the IT area was the pilot implementation of EHR (electronic health records) and EMR (electronic medical records). *HCwBQ* is aware of this being a difficult aspect to implement, and therefore they decided to adopt its implementation on a pilot basis. The management, however, feels bullish given the Indian government's push toward electronic governance and the current reforms in the tax system of the country. They consider it worth a try since the EHR-EMR implementations go well with the hospital's main objective of reducing carbon emissions, waste, and water consumption.

Although the dynamic CEO had much enthusiasm to embrace 'GREEN', it took a long time to convince the stakeholders to study the available material on greening of hospitals. Finally, in 2014 the hospital moved toward formalizing the launch of their "go-green" program. The program involved a number of phases. From healthcare privacy perspective, hospital's EHR implementation plan was based on the Electronic Health Records Standard for India published in August 2013 and approved by Ministry of Health and Family Welfare, Government of India. Figure 11.5 shows *HCwBQ*'s road map for the program based on the phases described.

The Phases (see Fig. 11.5)

1. Form the policy for "green strategy"
2. Prepare for executing the green strategy
3. Execute the strategy-based "greening approach" for the hospital
4. Monitor and measure the results
5. Maintain "green sustenance"

The Strategic Phase

During the policy-based strategy formation phase, the CEO led the study of energy usage in hospitals; it included the energy baseline and benchmarking in hospitals, the detailed study of energy efficiency opportunities in hospitals. During this study, the CEO found very useful the *Best Practice Guide* (for Energy Efficiency in Hospitals) published in March 2009 by USAID ECO-III Project. He also studied reports on the experience of EHR implementation in large hospitals. Among the other material he studied, were the various reports and guiding documents on energy efficiency in hospitals and energy efficiency in artificial lighting systems. His team studied the UNDP project document about "energy efficiency improvement in commercial buildings" – a program whose implementation partner was the Bureau of Energy Efficiency (BEE), Ministry of Power of Government of India.

Execution of the Green Strategy: A Quick Overview

1. Water Heating and Solar Energy Usage

The goal was to completely use solar energy for the hospital's internal requirements for water heating. With a large battery of solar panels on the roof of all buildings on the hospital complex, it was possible for *HCwBQ* to make a heavy use of solar energy (except during the monsoon season).

2. Energy Usage Optimization

HCwBQ adopted smaller measures such as switching over to compact fluorescent and light-emitting diode (LED) light bulbs, turning thermostats down by just a few degrees in the colder season or turning it up a little during the summers, purchasing energy-efficient lighting products, reducing standby energy use, and retrofitting buildings to cut energy waste. During the kickoff stage of the green implementation, the hospital launched an awareness campaign to reduce energy use throughout the hospital. The campaign included posters, slogans and other tools and measures such as training the staff to systematically turn off office equipment, reaping natural light during daylight hours in hospital corridors, and plugging leaks in the air conditioning. The project resulted in a total energy savings of 3,190,000 kWh from mid-2014 to the end of 2016 and a cost saving of INR 1,958,000.

3. Management of Waste

About 10–15% of the total waste generated in *HCwBQ* is hazardous waste. Knowing that such a waste can harm both its handlers and the environment, right from the beginning HCwBQ established systems for effective and safe disposal of their waste so that the waste generated would not find its way to the open bins on the roadside or the low-lying areas or be discharged into the water bodies. This was done to avoid the risk of improper disposal of such waste, especially as open dumps, attracting a host of disease vector, releasing unpleasant odors, and eventually leading to transmission of diseases.

HCwBQ reduces its solid waste and emissions through composting, recycling, simple green purchasing practices (minimizing packaging, using reusable rather than disposable products, and buying recycled products), and minimizing waste transport. The hospital has established contacts with appropriate agencies. Biomedical waste generated is handled through by utilizing the services of the chosen agency from among the CBWTF (Common Biomedical Waste Treatment Facility) from among the 188 operational ones. These agencies have 688 incinerators, 2710 autoclaves, 179 microwaves, 13 hydro-claves, and 4250 shredders as in-house treatment equipment. As far as the plastics are concerned, after carrying out disinfection, *HCwBQ* sends the used plastics for land filling (where appropriate and allowed with the permission from legal authorities), rather than getting it incinerated. The hospital does this dedicatedly keeping in mind that burning plastic produces greenhouse gases (GHGs) and toxic pollutants such as dioxins and furans. HCwBQ established a special plant to convert sewage into carbon-neutral renewable energy using a high-temperature thermophilic anaerobic digester. Bio-methane is naturally created by this high-tech compost system and is polished into bio-natural gas (BNG). The BNG is then used in a cogeneration plant to power the hospital and water purification systems.

4. A Futuristic Vision for Hospital's Energy Management Waste Management

There was a time when *HCwBQ* used to procure most of the energy from outside, either in the form of different fuels or electricity. Therefore, their energy bill constituted a significant proportion of their operational cost. Water pumps, air-conditioning and ventilation units, lighting, and innumerable other appliances, gadgets and devices were used directly or indirectly for the provision of patient care consume electricity. Processes such as catering, laundry, and instrument sterilization required steam. Furnace oil or gas (LPG)-fired boilers were used to generate steam. Diesel guzzling and smoke belching standby power generation sets are commonly employed for energy generation in hospitals across India, because of chronic power shortages. In view of the energy-poor settings, *HCwBQ* took the decision to harness low-energy and no-energy medical devices, together with the increase in the deployment of renewable energy (the solar panels on the rooftops of hospital buildings). Now they are using solar energy to sterilize medical equipment, sanitize the hospital's laundry, heat water for baths, light outdoor pathways at night, and power the residential building and office.

HCwBQ is aware of the fact that solid waste, though a large component of waste generated, is not the only form of waste. A sizeable amount of electronic and electrical waste is also generated on its campus (discarded batteries, bulbs, electric wires, electronic circuit boards from discarded PCs and monitors, and other electrical as well as electronic components, to name a few), although the quantity is not as much as that of solid waste. Therefore, *HCwBQ* has identified a vendor (selecting from among 77 registered eWaste recycling facilities) who specializes in recycling and environmentally friendly disposal of the eWaste. The hospital already uses types of bulbs that are less hazardous at the time of their disposal.

5. Reduction of Transportation Cost

A team of HCwBQ doctors and selected members of paramedical staff are now working together to investigate the potential of telemedicine for reducing air pollution, by reducing travel and transportation. Currently the hospital is in the process of procuring the required equipment to go ahead with the telemedicine pilot project later in the year. The primary objective is to improve the reach to rural areas of India. The possible reduction in patient commuting would be the secondary objective.

6. The Implementation of EHR (Electronic Health Record) System

The HCwBQ team, specially formed for EHR implementation, consisted mainly of the IT (information technology) staff of the hospital and doctors who have deep knowledge about patients' medical records. HCwBQ selected doctors who had high level of computer usage skill. Considering the overall skepticism about electronic records in general and EHR in particular, HCwBQ launched a systematic plan for creating EHR awareness among its patients. It took more than 8 months to get some visible awareness manifesting itself in the ground reality. Their experience was that patients with lower education level were not concerned with the EHR replacing the paper records, while the patients from more educated classes had far more concerns.

Other Visible Results

HCwBQ eliminated up to 53,000 tons of CO_2 emissions by replacing face-to-face patient visits with virtual visits. In addition, HCwBQ could avoid 5000 tons of CO_2 emissions by filing prescriptions online. The hospital could avoid the annual use of 1044 tons of paper for medical charts. They also reduced the use of toxic chemicals by 23.3 tons through the digitization and archival of X-ray images and other scans. Their green efforts resulted in a positive net effect on the environment despite increased energy use and additional waste from the increasing use of personal computers and tablets, iPads, etc. All this provided them one more benefit; HCwBQ became the leader among the Indian hospitals going green and also significantly improved their social image, which they publicized thru the use of active social media in the country. This in turn resulted in the improvement of the value of their stake in the share market.

Green Construction in HCwBQ's Expansion Plans

For its future construction plans, the management of the hospital is keenly exploring other ways of going environmentally green, such as the use of local and regional building materials, planting a large amount of trees on the hospital site, and incorporating design components like daylighting, natural ventilation, alternative energy, water harvesting, and green roofs. They want to consider rainwater harvesting in their future building. They aim to harvest rainwater during the monsoon and then store it for use during the dry months of the year, to recycle it, and to be able to use gray water for irrigation on hospital grounds to grow useful plants.

They have found out that the difference between the average costs for green buildings vis-a-vis non-green buildings is not significant. Furthermore, HCwBQ wants to continue the improvement of air quality in the indoor areas of the hospital. HCwBQ's research findings point to the relationship between improved indoor air quality and positive health impacts on illness, including asthma, flu, sick building syndrome, respiratory problems, and headaches. HCwBQ took much inspiration from the Sambhavna Trust Clinic, founded in 1996 in Bhopal, the Indian state of Madhya Pradesh. The said clinic is a green structure in every sense. It was constructed using local materials that are low cost and durable. It has combined beauty with function while blending in with the landscape and is passively cooled, day lit, harvests the rainwater, and uses solar water heaters.

Looking Back

HCwBQ's strong HIS (health information system) helped them in the EHR implementation. They worked with green-certified suppliers of building materials. They worked with experts in the domain of alternate sources of energy; the country being sunny almost throughout the year, helped maintain the system of solar panels that *HCwBQ* had right from its inception. They only had to bring in vendors of systems that would help them improve the storage of the solar energy. The management at *HCwBQ* learned that one of the greatest psychological barriers while going green is the mindset of its shareholders that creating healthy buildings costs more money; this, however, turned out to be a myth. Initially, the financial accounting team of the hospital was not keen on the green implementation given that traditionally, the accounting team likes to focus on the up-front costs. They have the tendency to disregard the essential "life cycle costs" of green products. However, the young CEO spent hours in convincing them about the long-term benefits of green implementation especially from the triple-bottom line perspective as well as from the social responsibility perspective. The CEO is convinced about the triple-bottom concept introduced and popularized by the green business guru John B. Ellington in the mid-1990s. *HCwBQ* is now looking to use other forms of renewable organic matter and geothermal heat.

Methodology Used for the Case Study

A case study is an empirical enquiry that investigates a contemporary phenomenon in depth and within its real-life context. We selected the said hospital (its real name masked as per the expressed requirement of the hospital's stakeholders) for a number of reasons – (1) it is a large hospital and has a strong reputation, (2) they were in the process of implementing the EMR system when contacted about the study, and at that time the hospital was still using paper and electronic records. This helped us to gain the understanding about the transition issues involved. Data for the case study was obtained during the period of sixty three days from April 12, 2017 to June 15, 2017. The data was collected through direct observations, semi-structured interviews, informal conversations, and the study of selected documents. Eighteen interviews were conducted with eleven actors (top management members, shareholders, doctors, members of IT staff and administrative staff, including the personnel in the medical records department). The pre-planned interviews were conducted face-to-face by the same person that conducted the observations. The interviews began with the interviewer introducing herself (the lead author of this book) and elucidating the purpose of the study. Confidentiality and privacy of the participant were guarded. The observations were conducted in three phases with 80 h effort expended. With management's permission, we also studied the necessary documents available at the hospital's web site as well as paper and electronic versions of the health records used at the hospital. Some additional follow-up information was obtained by email correspondence with our contacts at the hospital at the end of the formal data collection period.

This case study addresses the "why" and "how" of greening of a hospital. The goals of this case study (based on qualitative research approach) were (1) exploratory, (2) descriptive, and (3) explanatory. Given the nature of the investigation we undertook, based on author's extensive research experience in the green (IT) aspects of the healthcare domain, it was decided that the "case study research method" would be the best one for preparing this case study. It is also the best method to use when the research has little control over events (in this case, the dynamics of the chosen hospital for the case study). Thus, this green implementation illustration from real life is the result of a qualitative case study method used to collect the data. It also involves the use of other well-known research methodologies such as "structured interviews" and "expert panel judgment." As mentioned, for privacy reasons and the expressed agreement signed with the hospital to protect its identity, the real name of the hospital has been masked, while the case study presents the actual green implementation of the hospital. The overall approach used for the preparation of the case study is depicted in Fig. 11.6 below.

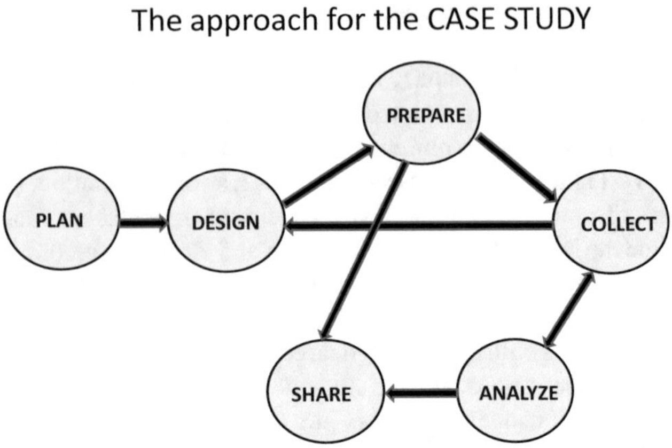

The approach for the CASE STUDY

Fig. 11.6 *HCwBQ* case study approach

Pointers to Other Chapters for Additional Details on Case Study Lessons Learned

This case study emphasized the need for collaboration, green IT, etc. Here are pointers to the chapters that give additional detailed information on those topics.

- Collaboration. Chapter 7
- Green IT. Chapters 5 and 7
- Water Heating and Solar Energy Use. Appendix A
- LED Lighting. Chapters 1 and 10
- Use of EMR and EHR. Chapters 2 and 3

India Case Study: Monalisa Healthcare Builds a Futuristic Data Center

The Background

Since 1990, Monalisa Healthcare Services (now a public limited company herein after referred to as 'Monalisa' in this case study) has been the leader in arranging the delivery of healthcare services to people eligible for government-sponsored programs such as "Health For All," "Affordable Medicare," and other similar ones. Through licensed health plan subsidiaries, approximately 70 million members are engaged in Monalisa's healthcare services programs spanning across ten Indian states. Monalisa Healthcare Services also provides primary care through clinics in the states of Punjab and Karnataka.

Well poised in its healthcare service delivery, Monalisa is now looking for growth and a strategy for the diversification of its IT (information technology) plan considering the new demands on the company's existing data centers. Monalisa's data centers deliver IT services to employees across approximately ten states. In order to meet the challenges of growth and to be ready for delivering new healthcare services in the future, Monalisa recently built a data center in Noida (in north of India).

In the first phase, they plan to consolidate into a single one, the operations from five of their data centers. The new data center encompasses 18,000 ft^2 and houses 250–300 physical machines, 600–700 operating instances, and 5 storage area networks.

The Phases

In the initial planning phases, the design and functionality of data center were influenced by a number of business priorities. First of all, Monalisa wanted to engage into collaboration among its employees across the organization. They chose Cisco equipment to build their company's main routing and switching infrastructure because of Cisco's leadership position in the domain of networking solutions. The company (i.e., Monalisa Healthcare) has planned to expand its current communications solutions to include new capabilities, such as instant messaging and web conferencing facilities.

Secondly, the data center was planned in such a way that it could attain gold certification from the LEED (Leadership in Energy and Environmental Design) green building rating system while supporting redundancy, business continuity objectives, and the company's anticipated growth. Finally, in the third phase, Monalisa planned to expand its virtualized server environment from 66% to 90% virtualization and to be ready for cloud computing solutions in the future. The CIO (chief information officer) Amar Devkar explains that the new data center of Monalisa Healthcare presents an opportunity to create a brand new network poised to maximize the latest capabilities of today's technology. He further explains that the new data center is instrumental to supporting company's growth without having to continually retrofit their data centers.

Although Monalisa's IT team is highly accomplished, the company wanted to seek technical assistance for the implementation of its new data center. The assistance sought included tasks such as managing the migration of their healthcare data and making the new data center fully operational. Monalisa contracted with their implementation partner for their end-to-end technological assistance in the achievement of the said objectives (from data migration stage through the operational phase of the new data center).

The management is satisfied to have chosen an implementation partner who understands the technology and the business processes that are critical to Monalisa Healthcare. They believe in their conviction that bringing together the right set of products with the right set of services can lead to meaningful improvements in healthcare environment. Their implementation partner helped them define an architectural vision for Monalisa Healthcare that is critical to the entire operation. The implementation partner provided them with not only guidance but also helped them implement best practices to ensure a successful migration. Lokesh Sanghvi, director of data-center operations for Monalisa Healthcare, explains that what contributed to the successful implementation of the new data center is the ability of their implementation partner who helped in ensuring that Monalisa was able to maximize the capabilities of their healthcare solutions to add value to their services.

The Approach

The Monalisa team worked with their chosen implementation partner to create an architecture needed for their current and future needs. Through the Data-Center Planning and Design Service, they assessed the project requirements, application environment, and business requirements. Next, a high-level design and BoM (bill of materials) were prepared for the new IT infrastructure. During the migration phase of the project, physical infrastructure, applications, and data were successfully migrated. In the next step, Monalisa's voice infrastructure and applications were shifted to the new facility. The implementation partner's information security expert provided assistance in creating a strong information security design as well as data privacy considerations for the protection of PHI (protected health information). Security design addressed authentication and authorization, data integrity, forensics, and security policy and compliance, and VPN (Virtual Private Network) requirements.

As Monalisa Healthcare moves forward with its server virtualization and collaboration plans, work is in progress to continuously improve network performance, prepare the network for change, and help the IT team succeed with new technologies. Monalisa management has gone for all the state-of-the-art technology latest-generation switches and unified computing system which play essential roles in their new data center. The unified computing system enables Monalisa Healthcare to easily scale its data-center capacity to meet their changing needs. The selected switches provide the high-capacity backbone of Monalisa's virtualized infrastructure significantly simplifying system management.

The network architecture is flexible enough to connect servers with either 1-Gigabit Ethernet (GigE) or 10-GigE for the deployment of Fibre Channel over Ethernet (FCoE) connectivity in the future. One operating system and a single point of management through the 5000 switches simplify provisioning, deliver operational consistency and transparency, significantly reduce the time required to add network capacity, and reduce the cost of cabling infrastructure. A single 10-GigE connection between servers and the computing platform supported Monalisa's virtualization strategy and greatly reduces the costs associated with ports, cables, adapters, and cooling requirements.

Achievements

The implementation partner helped Monalisa Healthcare implement its new data center on time and under budget and delivered right-sized capabilities for its needs. They consolidated multiple centers to extend its virtualization initiative, and they simplified the management of their healthcare data. The objective for Monalisa management was to significantly reduce its circuit, facility, and energy costs. They are confident that having the right footprint on the network, storage, and server side will allow them to have a lower cost structure and be more competitive. Monalisa anticipates a cost savings to the tune of 20% through the use of unified computing architecture. The new data center provides a foundation for supporting Monalisa's growth over the next 5 years. With the state-of-the art switches and the use of unified computing systems at its backbone, Monalisa's infrastructure will enable them to reach its 90% virtualization goal and reduce operations costs while maximizing application performance. The new IT and networking architecture will also provide a transparent environment for scaling operations, allowing Monalisa Healthcare to easily add new product lines, deliver new constituent services, or integrate company acquisitions without having to redesign the network. The unified architecture will simplify management of all of Monalisa's resources. The new data center also supports Monalisa's healthcare service collaboration initiatives. Together, these solutions create an integrated collaboration platform that enables employees to communicate more easily. After implementing collaboration technologies, Monalisa's staff finds that the patient treatment related discussion meetings have become more efficient. The teleconferencing systems implemented helping the delivery of healthcare services through telemedicine practices.

Time and costs associated with travel is reduced considerably. The CIO Amar Devkar estimates that the collaboration tools have saved a huge amount of travel costs and at the same time, productivity has been enhanced. Tina Baheti, Monalisa's chief operating officer says that as an organization, they committed to green initiatives. She further adds that their telemedicine initiatives are an integral part of their green strategy. Monalisa Healthcare can minimize the negative impact of their patients' travel related carbon footprint through the use of their unified communications and teleconferencing tools.

Looking Forward

Monalisa Healthcare looks forward to continue working with their implementation partner to achieve their goals. The CIO (Amar Devkar) says that their implementation partner took a holistic look at IT environment of Monalisa Healthcare, and they were able to design an architecture that will stand in good stead for many years to come.

The entity Monalisa Healthcare Ltd. Industry: Healthcare Location: India	Challenges: • The need to deploy data center to support business transformation • The pressure to achive rapid growth and increase efficiency • To Expand collaboration capabilities efficiently
Implementation: • Parnter company's full involvement in planning, design, data center migration, security, and implementation • State-of-the-art switches and Unified Computing System • Collaboration solutions	Achievements: • Reduced costs: data center circuit, energy, and facility • Simplification of data center management • On track to complete consolidation and migration on time and within budget

Fig. 11.7 Summary of the Monalisa case study

Pointers to Other Chapters for Additional Details on Case Study Lessons Learned

This case study emphasized the need for collaboration, green IT, etc. Here are pointers to the chapters that give additional detailed information on those topics.

- Collaboration. Chapter 7
- Green Healthcare, Green IT. Chapters 5 and 7
- Telemedicine. Chapter 3
- Network Design and Data-Center Efficiency. Chapters 3 and 5
- LEED. Chapter 9

Summary of the Case

Figure 11.7 presents the summary of the case in terms of the entity context, the challenges they had, and a short summary of the implementation of the data center and their achievements.

Chapter Summary and Conclusions

As discussed at the beginning of this chapter, the challenges in implementing green healthcare and green hospitals can vary considerably based on the part of the world where the healthcare facilities are located. Challenges include government regulations, the cost of electricity, and the social/political environment relating to the environment and energy efficiency. The first few chapters discussed the various aspects of green healthcare. The case studies in this chapter included green healthcare for countries around the world, including countries in Europe, Asia, Africa, and North and South America. Case studies for hospitals in the United States were discussed in Chap. 10. Although the social and political environments of the particular country do have an impact on implementing green healthcare, the technical challenges remain the same. The significant value in looking at green healthcare case studies should be to take advantage of lessons learned.

This chapter's worldwide case studies indicated that green healthcare challenges are very much the same all over the world. Thus collaboration between green healthcare groups worldwide is very important and will continue to grow in importance as new ideas are generated.

Additional Reading

Muduli, K., & Barve, A. (2012). Barriers to green practices in healthcare waste sector: An Indian perspective. *International Journal of Environmental Science and Development, 3*(4).

Chapter 12
The Future for Making Healthcare Green

What's past is prologue.

—William Shakespeare, The Tempest

*We're running the most dangerous experiment in history right
now, which is to see how much carbon dioxide the atmosphere...
can handle before there is an environmental catastrophe.*

—Elon Musk

The future of green healthcare is being shaped now – based on university, corporate, and government research, regulations, incentives, etc. Because of the dynamic aspects of green healthcare and green IT, we have already made significant progress with green healthcare and can build on that past. This chapter gives a summary of what we have discussed over the past 11 chapters and takes a look at the future of green healthcare and green IT.

For success with green healthcare, it needs to continue to be a very collaborative endeavor. The IT hardware and software manufacturers (Dell, HP, IBM, Fujitsu-Siemens, Intel, EMC, Microsoft, etc.) have a great deal of competitive incentive to market and improve the energy-efficiency aspects of their products. Energy utilities, government regulators, environmental advocates, and the groups involved in helping build data centers – including the infrastructure technology providers – all need to collaborate with IT providers and IT customers.

One of the most important arenas for development of green IT is that we must have better measurements in order to better manage energy use at data centers and throughout the healthcare enterprise. We will see great strides here, for example, gas mileage monitoring on the Honda Prius hybrid. Currently we can use the data network "sniffer" concept to determine network bandwidth and response time for each user of a server. That same sniffer concept will be developed for power and energy used. There will be an inexpensive power/energy monitoring appliance that will act like the network sniffer and give us information on the power and energy use for each IT device. That appliance will send the information to an energy monitoring system (a server). Then we can see actual energy reduction results of using virtualization and other energy saving initiatives. The electric power/energy monitoring device will also be in our homes and will have the capability of sending information to our laptops. Then we'll be able to see the actual power and energy-use history of our refrigerators or window air conditioners. We then will be able to better manage our energy use by measurements and trending.

Review of Green Healthcare Six Sigma Steps

According to most research, the "green wave" continues to rise for green healthcare. Currently, most IT organizations place environmental concerns within their top six buying criteria, and most companies (including healthcare companies) use carbon-footprint considerations with their hardware-buying strategy. Large enterprises continue to develop policies requiring their suppliers to prove their "green credentials" through an auditing process.

Of course, some companies are talking a good game but are not being as aggressive in going green as will be necessary to make the difference needed to solve our energy and climate crisis. Enterprises that have started the green journey, however, have found that reducing total energy requirements can be accomplished through some fairly straightforward improvements that don't take years to implement or bring return on their investment.

In Chaps. 2 and 4, we discussed the Six Sigma steps that apply to all companies and organizations in going to green healthcare and green IT. These steps, which have been discussed throughout the book, are:

1. Define the Opportunities and Problems (Diagnose)

Communicating to all employees your organizations plans and goals to save energy via green healthcare and Green IT is a very important step. That communication, in addition to plans and goals, needs to include the organization that will be driving the effort. It is a good idea to designate a focal point, with a title. The title doesn't need to be officially "energy czar" but in essence that should be the role of the focal point. Chapter 4 further discusses these concepts.

2. Measure (You Can't Manage What You Can't Measure)

A first step is to be able to measure current energy use and establish a baseline. The significance of consolidating and virtualizing IT is the topic of several chapters. Consolidation of IT operations and using virtualization to reduce server footprint and energy use are the most well-recognized and most-often-implemented effi-ciency strategies of the past few years. The checklist in Appendix A provides a good way to start with an inventory of where you are with green IT, and this checklist emphasizes consolidation and virtualization as key steps.

3. Analyze IT Infrastructure with Energy-Efficient Cooling

Chapter 5 describes the significance of energy-efficient cooling (cooling accounts for half of data-center energy use). Although energy-efficient data-center cooling is not something we IT people usually are involved with, the significance of energy-efficient data center needs to be appreciated as part of your green IT plan.

4. Improve (Virtualize IT Devices: Use Cloud When Feasible)

Chapters 7 and 8 describe the aspects of healthcare energy use and carbon-footprint metrics and the need for collaboration with groups such as The Green Grid. Over the past few years, The Green Grid has grown from 11 founding members to a consortium of more than 150 companies working to improve data-center energy efficiency. The group continues to release some of its most important deliverables in the form of metrics that businesses will be able to use to measure the power usage effectiveness of facilities infrastructure equipment. Most businesses can already readily identify areas where infrastructure optimization can achieve increased efficiency by simply monitoring and measuring their existing infrastructure equipment. Additionally, the Environmental Protection Agency is stepping in to help create metrics as well. About 100 companies have indicated they will provide raw power data and other information to the EPA for use in developing its new benchmark.

All companies and healthcare institutions should continue to make sure the utility costs associated with their data-center operations are broken out separately from those for other corporate facilities. In addition, metering specific equipment racks or types of equipment such as servers can provide valuable insight into which specific consolidation, virtualization, and optimization projects would yield the best ROI going forward.

Energy optimization software is discussed in Chap. 4 in the section "Six Sigma Approach to Help Make Healthcare Green." This discusses energy management and assessment software. Appendix A further discusses available energy monitoring and management software tools.

5. Control (Continue to Measure and Manage and Upgrade as Necessary)

Chapter 6 discusses the significance of software and application and storage efficiency for green IT. Data-storage efficiency such as the use of tiered storage is also very significant for reducing energy use. Data deduplication (often called "intelligent compression" or "single-instance storage") is a method of reducing storage needs by eliminating redundant data. Only one unique instance of the data is actually retained on storage media, such as disk or tape. Redundant data are replaced with a pointer to the unique data copy. For example, a typical email system might contain 100 instances of the same 1 megabyte (MB) file attachment. If the email platform is backed up or archived, all 100 instances are saved, requiring 100 MB storage space. With data deduplication, only one instance of the attachment is actually stored; each subsequent instance is just referenced back to the one saved copy. In this example, a 100 MB storage demand could be reduced to only 1 MB.

Chapters 4 and 5 describe the many incentives that encourage businesses to update equipment and adopt efficient operational practices. These practices can help reduce peak and total power demands and that's what makes the electric power utilities happy.

Chapter 7 described how going to green IT on a worldwide basis requires the collaboration of many different groups, including electric utilities, government agencies, IT technology vendors, data-center design and build businesses, and all companies and organizations worldwide, including your company. That's one of the best things about green IT as a step toward solving the energy crisis – we can all participate. The case studies in Chaps. 9, 10, and 11 and the Appendices give details on how other companies have implemented green IT.

Green Healthcare for the Future

Green computing promises an enormous win for IT: a chance to save money and the environment. Many companies and hospitals are trying to go greener, but a few truly stand out. Many companies have begun to address some of these challenges head on. In fact, 2017 information indicates that almost all large enterprises have consolidated or started to consolidate IT systems, and most are doing some level of virtualization. Those that have really advanced these efforts are seeing significant returns or savings. Some customers have shown the ability to:

- Triple asset utilization
- Provision new resources in minutes
- Reduce heat by up to 60%
- Reduce floor space by as much as 80%
- Reduce disaster recovery time by 85%

The following section describes one vision for a green IT and green healthcare future.

A Road Map for Green Healthcare

Many of the ideas in this section are based on the authors' work with IBM over many years to create more energy-efficient data centers (which includes data centers for hospitals and other healthcare facilities).

You can't make the world move more slowly. Or change where markets are headed. Or hold back new technologies while focusing on day-to-day IT operational issues. But there is something you can do, right now.

The fact is, not all of today's IT infrastructures were built to support the explosive growth in compute capacity and information. Many data centers have become highly distributed and somewhat fragmented. As a result, they are limited in their ability to change quickly and support the integration of new types of technologies or to easily scale to power the business as needed. So how do you find the time and resources to drive the innovation required to keep your healthcare facility competitive in a rapidly changing marketplace? How can you react to business needs faster?

Since today's distributed approach to the enterprise data center is challenged to keep up in a fast-paced business environment, a new centralized IT approach is needed. We must rethink IT service delivery to help move beyond today's operational challenges to a new data-center model that is more efficient, service oriented, and responsive to business needs.

This vision for the new enterprise data center is an evolutionary model that helps reset the economics of IT and can dramatically improve operational efficiency. It also can help reduce and control rising costs and improve provisioning speed and data-center security and resiliency – at any scale. It will allow you to be highly responsive to any user need. And it aligns technology and business – giving you the freedom and the tools you need to innovate – and stay ahead of the competition.

Through our experience with thousands of client engagements, we have developed an architected approach based on best practices and proven implementation patterns and blueprints. And our own data-center transformation provides first-hand proof that embracing this new approach simply makes good business sense.

Right now, technology leaders are challenged to manage sprawling, complex distributed infrastructures and an ever-growing tidal wave of data while remaining highly responsive to business demands. And, they must evaluate and decide when and how to adopt a multitude of innovations that will keep their companies and healthcare organizations competitive. IT professionals spend much of the day fixing problems – keeping them from applying time and resources to development activities that could truly drive business innovation. In fact, many say they spend too much time mired down in operations and precious little time helping the business grow. These operational issues include:

- *Costs and service delivery:* Time is money – and most IT departments are forced to stretch both. There is no question that the daily expense of managing operations is increasing, as is the cost and availability of skilled labor. In fact, IT system administration costs have grown fourfold, and power and cooling costs have risen eightfold since 1996 alone. And in today's data center, data volumes and network bandwidth consumed are doubling every 18 months with devices accessing data over networks doubling every 2.5 years.

- *Business resiliency and security:* As enterprises expand globally, organizations are requiring that IT groups strengthen the security measures they put in place to protect critical information, for good reason. Enterprise risk management is now being integrated into corporate ratings delivered by organizations such as Fitch, Moody's, and Standard & Poor's. At the same time, companies are demanding that users have real-time access to this information, putting extra – and often conflicting – pressure on the enterprise to be both secure and resilient in the expanding IT environment.

- *Energy requirements:* As IT grows, enterprises require greater power and cooling capacities. In fact, energy costs related to server sprawl alone may rise from less than 10% to 30% of IT budgets in the coming years. These trends are forcing technology organizations to become more energy efficient – to control costs while developing a flexible foundation from which to scale.

The bottom line is that enterprises report that IT operational overhead is reaching up to 70% of the overall IT budget. And that number is growing – leaving precious few resources for new initiatives.

If you're spending most of your time mired in day-to-day operations, it's difficult to evaluate and leverage new technologies available that could streamline your IT operations and help keep your company competitive and profitable. Yet the rate of technology adoption around us is moving at breakneck speed, and much of it is disrupting the infrastructure status quo. Consider some examples: in 2017, it is estimated that medical imaging consumes more than 30% of the world's data storage. Increasing speed and availability of network bandwidth is creating new opportunities to integrate services across the web and re-centralize distributed IT resources. Access to trusted information and real-time data and analytics will soon become basic expectations. Driven by the expanding processing power of multicore and specialty processor-based systems, supercomputing power will be available to the masses. And it will require systems, data, applications, and networks that are always available, secure, and resilient.

Further, the proliferation of data sources, RFID and mobile devices, unified communications, SOA, Web 2.0, and technologies like mashups and XML create opportunities for new types of business solutions. In fact, the advancements in technology that are driving change can be seen in the new emerging types of data centers, such as the Internet and Web 2.0, which are broadening the available options for connecting, securing, and managing business processes. Ultimately, all of these new innovations need to play an important role in the new enterprise data center. "More than

70 percent of the world's Global 1000 organizations will have to modify their data center facilities significantly during the next five years."

The vision for the new enterprise data center provides for a new approach to IT service delivery. Through it, you can leverage today's best practices and technology to better manage costs, improve operational performance and resiliency and quickly respond to business needs. Its goal is to deliver:

New economics: The new enterprise data center helps you transcend traditional operational issues to achieve new levels of efficiency, flexibility, and responsiveness. Through virtualization you can break the lock between your IT resources and business services – freeing you to exploit highly optimized systems and networks to improve efficiency and reduce overall costs.

Rapid service deployment: The ability to deliver quality service is critical to businesses of all sizes. Maintaining a positive customer experience – and ensuring cost efficiency and a fast ROI relies on your ability see and manage the business – while leveraging automation to drive efficiency and operational agility. Therefore, service management is a key element in the new enterprise data-center approach.

Business alignment: A highly efficient and shared infrastructure can allow you to respond instantaneously to new business needs. It creates opportunities to make sound decisions based on information obtained in real time, and it provides the tools you need to free up resources from more traditional operational demands. With a new enterprise data center, you can focus on delivering IT as a set of services aligned to the business – freeing up time to spend on IT-driven business innovation.

What makes this approach for efficient IT service delivery so unique? As businesses move toward a recentralization of the data-center environment, a holistic integrated approach needs to be considered. We need to capture an end-to-end view of the IT data center and its key components. Although we understand that incremental improvements to each element of the new enterprise data center can improve overall operations, we take into account that modifications to one component may strain the performance of another.

For example, upgrading the enterprise information architecture to provide integrated and trusted information to users will likely require changes to security and business resiliency approaches. And, creating highly virtualized resources are most effective along with a stronger, more integrated service management approach. As such, the strategy for the new enterprise data center needs to be holistic and integrate the key elements of:

- *Highly virtualized resources* that are flexible to adjust to changing business needs to allow for more responsive provisioning and help deliver efficient resource utilization. Virtualization removes the bind between applications and data and underlying physical resources – granting IT organizations more flexibility and freedom in deployment options and the ability to exploit highly optimized systems.

- *Business-driven service management*, in which a complex and difficult-to-manage environment is transformed for improved transparency – and cost-efficient, easier management. This transformation involves raising management tasks from the simple monitoring of individual resources to the orchestration of the entire environment to be more responsive and efficient. Once transformed, the environment can be fully aligned with business needs and controls to ensure that customer priorities are met, business controls are maintained, and availability and performance are maximized across the entire enterprise.

- *Security and business resilience approaches* and best practices that become increasingly important with the consolidation of data centers and recentralization of systems and data while providing secure, open access across and beyond organizational boundaries.

- *Efficient, green, and optimized infrastructures and facilities*, which balance and adjust the workloads across a virtualized infrastructure and align the power and cooling consumption with business processing requirements across all IT and data-center facilities. The result is balanced energy demands to help avoid high peak energy use and the associated higher energy billing rates and meet SLAs based on business priorities. Through the introduction of an optimized infrastructure, the number of systems and networks in the data center can be reduced, cost efficiency improved, and energy efficiency enhanced.

- *Enterprise information architecture.* Data that was typically contained in disconnected, heterogeneous sources and content silos is virtualized through a flexible enterprise information architecture. Therefore, IT can deliver trusted information to people, processes and applications to truly optimize the business decision-making and performance.

Conclusions

As we've seen throughout this book, green healthcare and green IT promise a significant win for healthcare and IT: a chance to save money and the environment. Collaboration with governments on all levels, energy research organizations, universities, energy utilities, IT vendors, and all of the nonprofit green organizations that keep springing up is key. And it's not just for organizations dealing with IT. Almost everyone worldwide can collaborate on green healthcare and green IT, since everyone is a user of healthcare and almost everyone is a user of IT through PCs, the Internet, and cell phones. We all need to contribute to energy efficiency to help solve the climate crisis. Energy conservation will remain the best and easiest way to save energy – and, of course, that applies to energy conservation for our IT systems and for every other energy-consuming device we use.

Additional Reading

Anusha, S.K., et al. (2016). Overview of big data's contributions in health care. *Imperial Journal of Interdisciplinary Research (IJIR) 2*,(11).

Ebbers, M., et al. (2008). The green datacenter: Steps for the journey. *IBM Redpaper*, http://www.redbooks.ibm.com.

Emerson/Liebert white paper. (2008) Five strategies for cutting data center energy costs through enhanced cooling efficiency. Emerson/Liebert, http://www.energyefficientdatacenters.techweb.com/login.jhtml?_requestid=882321.

Esty, D.C., & Winston, A.S. (2006). *Green to gold: How smart companies use environmental strategy to innovate, create value, and build competitive advantage.* Yale University Press. 384 pages, 978-0300119978.

Hiremane, R. (2008). Using utility rebates to minimize energy costs in the data center. *The Data Center Journal*, http://datacenterjournal.com/index.php?option=com_content&task=view&id=1475&Itemid=41 February.

Mitchell, R. (2007). Seven steps to a green data center. Computerworld.

Molloy, C. (2008). Project big green: The rest of the story. *IBM Technical Leadership Conference (TLE)*, Orlando.

Pacific Gas & Electric (PG&E). (2008). Data center energy management. http://hightech.lbl.gov/DCTraining/.

Comments: This web site includes tools and information developed with the Lawrence Berkeley National Lab (LBL) that will prove useful to help you determine savings opportunities for data centers.

Pacific Gas & Electric (PG&E) (2006). High performance data centers. http://hightech.lbl.gov/documents/DATA_CENTERS/06_DataCenters-PGE.pdf, January.

Patterson, M.K., Costello, D., Grimm P, Loeffler, M. (2007). Data center TCO; a comparison of high-density and Low-density Spaces. THERMES 2007, Santa Fe.

Rassmussen, N. (2005). Electrical efficiency modeling of data centers. White Paper #113, APC. http://www.apcmedia.com/salestools/NRAN-66CK3D_R1_EN.pdf.

Rodriguez, J-M. (2008). Green data center of the future – architecture. IBM Technical Leadership Conference (TLE), Orlando, Florida.

Velte, T.J., Velte, A.T., Elscnpeter, R. (2008). *Green IT: Reduce your information system's environmental impact while adding to the bottom line.* McGraw-Hill. ISBN: 978-0-07-159923-8.

Appendix A. Green IT Checklist and Tools and Calculations for Healthcare Energy and Carbon Footprint Estimates

This appendix gives information on how to estimate healthcare server annual power and cooling usage and costs. Included would be methods for estimating the environmental impact (e.g., emissions) and economic impact (e.g., monthly costs) for different types of servers, data storage, etc.

> ... you can't make a product greener, whether it's a car, a refrigerator or a traffic system, without making it smarter – smarter materials, smarter software or smarter design.
> —Thomas L. Friedman – Author of "Hot, Flat, and Crowded"

This appendix includes a green IT checklist and information on various green healthcare and green IT areas.

A Green IT Checklist

The "green IT checklist" given in this section was developed by David F. Anderson PE, who has been a Green Architect for IBM's Energy Efficiency Initiative and in 2018 supports IBM's IT Optimization Team which is focused on helping organizations save money and energy. One of David's passions has been to assist companies (large and small), schools, and other organizations conducting studies for IT optimization and giving talks on green IT. David's background as an engineer includes Professional Engineer (PE) certification, Industrial Engineering, Electrical Engineering, and Civil Engineering giving him a diverse but necessary background to address all aspects of green IT. This checklist was developed as a result of David's speaking engagements and hundreds of interactions with organizations focused on green IT and saving money through optimizing IT.

© Springer International Publishing AG, part of Springer Nature 2018
N. S. Godbole, J. P. Lamb, *Making Healthcare Green*,
https://doi.org/10.1007/978-3-319-79069-5

Tips on Creating Sustainable Data Centers (David Anderson PE)

Everywhere I go from the grocery store to the car pool parking place outside IBM's mega data centers, I am now reminded to be green. Whether it be purchasing alternative "green" products for everyday use or taking care of business in a sustainable way, we now have the opportunity to change for the better. I simply define better as doing operations with less energy and creating longer lasting value. Green is all about doing the right thing *and* saving costs. How can you optimize your data center and IT operations to save energy and costs? Here are some tips and examples of *green IT practitioners.*

1. *Begin with an Enterprise Goal in Mind: Create Lasting Greenness*

Small scale will yield small results. Large scale will yield bigger results. Often every project looks good and green by itself, but when added up, a suboptimal data center and complex infrastructure has been created. IBM has a vision that can become a green blueprint for a data center that has state-of-the-art capabilities while using less energy and space. The state-of-the-art data center exploits virtualization, uses service management with automation, can run public and private clouds, and aligns with business goals. The journey of transformation includes being simplified, shared, and dynamic. Start a green program with goals in mind. Design and architect in qualities that are required. Security, reliability, availability, serviceability (RAS), scalability, and flexibility are "table stakes" or characteristics that are a must-have for almost all organizations. Adding capacity in increments when needed enables the most efficient use of physical and energy-consuming assets. Underutilized assets are wasteful from both energy and resource perspectives. Would you allow your people to only work 5–10% of the time? Why have servers and their support infrastructure that are only 5–10% utilized?

2. *Exploit Virtualization to Reduce the Number of Servers and Improve Flexibility*

Virtualization or using software/hypervisor technology to represent virtual servers rather than physical servers is a very green technology. Virtualizing physical servers has become the norm for the vast majority of server assets. Reducing the number of power drawing components in the data center to a minimum directly slashes the amount of energy consumed, reduces the cooling requirements, and can substantially impact (reduce) the software licensing requirements associated with running operating systems, middleware (such as databases and web and application server), and workloads. Logical partitioning (LPAR) and virtualization technologies such as VMware, PowerVM, and z/VM, to name a few, break the physical boundaries of servers and drive up utilization, reducing the need for as many servers. Starting with pilot or proof of concept projects is easier than ever, since IT vendor services and virtualization technologies on all platforms have matured throughout the twenty-first century. From open-source virtualization to mainframe z/VM virtualization, implementation services and technologies abound to start eliminating the wasteful approaches of one server per application and a variety of nonproduction servers (development, test, quality assurance, unit assurance testing) for every production server. Many applications can be hosted on a physical server and still have autonomy because of virtualization. Virtualization allows servers to be "spun up" and collapsed quickly moving capacity to applications where needed. How many servers can be eliminated by virtualizing? Compression ratios of 1–8 are common, and 1–50 compression ratios are often achieved with the best virtualization technologies such as IBM's z/VM which is now over 50 years old and still evolving to add value. VMWare is the most common virtualization technology used for X86 servers. PowerVM is the most common technology used for Unix servers.

3. *Exploit Virtualization to Reduce the Amount of Storage and Networking Equipment*

This includes SANs and using virtual I/O connections within servers such as IBM's HiperSockets, Virtual Ethernet for Power and IBM i Systems, and OSA integrated layer 2 and 3 switching. Virtualization can be done for not only all three subsystems of a server (processor, memory, and I/O) but can be used in the connection of servers to other servers, the world outside the data center and data. Data is the lifeblood of organizations. Storage devices virtualize to optimally use the assets. Storage controllers have improved in the twenty-first century as storage densities and costs have substantially decreased.

4. *Use Integrated Approaches to Server Consolidation to Optimize Savings*

More than a single methodology needs to be applied to get the fewest number of servers. IBM's enterprise computing model for reducing thousands to about 30 large centralized servers used the following approach to consolidation:

- Migrate servers delivering largest savings first (i.e., stranded infrastructure or costly software). This primes the pump and generates enthusiasm and savings for other green projects.
- Eliminate assets with lowest utilization first. These assets are not pulling their weight when measured by watts/virtual or logical image or other common metrics to compare servers and normally are older and less efficient than newer technologies.
- Identify assets with an upcoming compelling event to mitigate expense (upgrade, move, and/or asset refresh). It is always easier to have a positive ROI and be green within the normal refresh period of assets.
- Aggregate by work portfolio to leverage strong customer buy-in. Ease of migration assists speed and successful workload migrations. People costs related to migrations can be substantial and should be minimized where possible.
- Start with oldest and least capable technologies first as they use the most power and provides the least performance per virtual or physical workload/ application.
- Focus on freeing up contiguous raised floor space. This enables growth and the addition of energy-efficient new IT and facilities equipment.
- Provision of new applications to an enterprise-class large centralized server. Mainframes, large IBM power servers, and X86 servers with many sockets and cores can efficiently run dozens or hundreds of workloads.

5. *Drive to High Utilization Rates for Private and Public Clouds*

Virtualization and management of workloads is key. The operating system and/ or cloud provisioning software must manage workloads to business priorities and dispatch in an automated manner. The average Wintel server is used only 5–20% of the time. No managers would allow their people to work 5–20% of the time. With new technologies and automation, utilization rates can go beyond 50% and at the same time improve flexibility and responsiveness as more resources can be tapped for peaks while better utilizing physical assets to reduce energy consumption.

6. *Consolidate on Large Servers*

Fewer larger servers will convert AC to DC more efficiently than many smaller servers with smaller and less efficient power systems. Power supplies of large servers are capable of operating at very high efficiencies (+90%). Large servers can also take advantage of high voltages and eliminate a conversion loss that robs efficiencies when stepping down to smaller voltages. The ability to more efficiently share resources makes running on a few larger systems more efficient than many small ones. Workload can be balanced, driving up utilization and reducing the number of watts needed to run applications, day or night.

7. *Eliminate Redundancy But Keep High Availability and Disaster Recovery Capabilities*

High availability and disaster recovery can be efficient and in a green way be designed into server configurations. Engines can now add non-disruptively to almost all platforms, reducing the need for extra servers. No longer is an idle server needed for *what if* scenarios. Production servers can back up other production servers. Configuring the ability to non-disruptively add (and reduce) capacity for production or disaster recovery without having idle or underutilized servers significantly reduces the number of footprints and slashes the energy consumed in the data center. Commonly used technologies include IBM's On/Off Capacity on Demand (add engines by the day) and Capacity Backup Upgrade (CBU) for disaster recovery. A data center can be greened and the bottom line improved by using fewer servers while having the ability to increase capacity without adding server and the associated facility infrastructure. Balancing between private and public clouds can also improve asset utilizations in various data centers.

8. *Measure and Put the Costs of Energy Where They Are Incurred*

Automated measuring and billing of energy consumption makes usage part of cost and green decisions. Without energy and cooling knowledge, requirements are unknown, inaccurate, and often over-planned leading to inefficiencies. An example of new technology to optimize energy use is IBM's Active Energy Manager (AEM). Monitoring energy usage and developing trends is key to understanding how energy is being used. This first step to optimizing energy use opens up the potential to become more efficient and optimizing for performance/watt. Managing energy use is an evolving concept in the data center. Capping power at the server level and optimizing to deliver the right performance per watt can be achieved using AEM. In the future the most efficient data centers will treat servers like you treat lights in your house, turning them off when not in use or at least turning on only what you need. Linkage to total cost of ownership (TCO) ensures green is part of every decision. Benchmark the entire data center as well as local areas for continuous improvement. Use commonly accepted methodologies such as the power usage efficiency ratio (total power/IT equipment power = PUE) from the Green Grid Consortium or the energy efficiency ratio. If local subject matter experts to assist optimizing are not available, use industry consultants such as IBM's IT Economic Optimization team.

9. *Use the Concept of Hierarchical Storage*

Picking the right media and format for storing data is like picking the right vehicle for a trip. Not every trip needs an 18 wheeler or a motorcycle. A combination of disk, tape, and hybrid technologies optimizes the use of energy while giving your data a secure and extendable home. Tape, a green storage equipment star, uses the least amount of energy and should be part of the storage constellation. Disk storage should be for demanding applications that require frequent updates. The Virtual Tape Server can mask latency with many applications and is another green star in the storage constellation. Larger and slower disks use less energy, and if their latency can be masked, the energy efficiencies gained by their use are worth it. For less demanding apps, MAID may be appropriate, and the elimination of spinning disks when not needed can substantially reduce wasted watts.

10. *Use the Latest Equipment*

Newer generations of IT equipment are more energy efficient and give better performance than older IT equipment. Begin greening your data center by replacing *the oldest, most inefficient equipment first*. Newer generations of servers and storage are built with more efficient power supplies, processors, memory, and I/O. Just about everything in newer servers and storage provides more performance and stores more data with fewer watts. We all have experienced how digital cameras have provided more memory, function, and better performance in the last decade. Servers and storage are on similar technology improvement trajectories. Servers scale higher in performance while using fewer watts per logical image. Decommissioning older servers that were never designed for virtualization or energy efficiency can be one of the most cost effective ways to green your data center. Like the gas-guzzling clunker that needs to be replaced with a hybrid, there are better ways now to run applications. There is a *BIG* difference between IT and your automobile. The new servers allow you to replace many older servers. The simplicity in running fewer physical servers is a no-brainer. The System z14, IBM's new generation of energy-efficient large centralized servers, can replace over 5000 older distributed servers. The energy efficiency and space savings enable the data center to add capacity within the existing four walls.

11. *Do Not Wait for the 11th Hour: Start to a Sustainable Data Center Now*

The biggest savings of all for going green is to create a culture and infrastructure that exploits technology for creating a sustainable data center. With green concepts and projects, a data center can grow in capability/capacity while continuing to use the same or less space and energy. Conducting or having a third party conduct an energy audit can benchmark where you are and identify projects with ROIs that can be prioritized to give cascading green returns. For every watt you save in IT equipment, you reduce the infrastructure (UPS, cooling, etc.) load and generate savings for future projects. Conserving energy in the data center allows the dollars to be used for adding more value to the business. Use energy like a precious commodity. Turn it up/on when needed and throttle back/turn off when not needed. Create a culture and data center that is intrinsically green.

Additional Green Ideas

The following are some additional green ideas for sprucing up your data center. Some you may have already done. Others may yield small to massive energy savings. All of them I have observed in various data centers in the last 10 years.

- Know the latest ranges of the temperature and humidity specs of ASHRAE. Stop running too hot or cold. Yes, you can run your data center between 60 and 80 °F. Let the hot aisles be hot.
- Place equipment so that it is in hot and cold aisles with two-floor tile width. Stop mixing hot and cold air wherever possible.
- This includes keeping openings in the server ranks and the servers to a minimum. Inspect each cabinet making sure that for all empty slot positions where no equipment is installed, filler strips or blanking plates are installed to eliminate turbulence inside the cabinet (allowing proper cooling of the installed hardware). In some cases, where a cabinet is by itself, devices such as "snorkels" may be used to direct either cold air into or hot air out of the cabinet as a tactic to provide the most efficient cooling for that cabinet. This is especially useful where a hot side of a cabinet faces a cold side of an adjacent cabinet.
- If you need "enclosures" for additional cooling, you still should have two-floor tile width for servicing. One is tight and three is wasteful.
- Do an assessment that looks at both temperature and airflow. IBM has advanced tools and models to assist this "baselining" of what the cooling profile is.
- Use free cooling. Outside air can substantially reduce energy required by computer room air conditioners. Data-center site selection will enable more days of free cooling when climate has big difference between day and night temperatures. Colorado is an example of an excellent climate to exploit free cooling.
- Enable active energy management (IBM Power Director Active Energy Management with Tivoli). If you do not measure, you miss understanding easy

opportunities to improve. Measuring enables new charging methodologies that include energy consumption.

- Charge for usage and have a surcharge for equipment at peak demand (or a reduced rate at off hours).
- Larger than code copper distribution (wiring for data center).
- Wherever possible *eliminate* conversion losses. This included using high-voltage power as well as rotary UPS.
- Plan for use of 480 V (or 600 V) to equipment. The mainframe uses this today.
- Enable dynamic provisioning of server and storage resources. This can reduce as well as turn off number of servers drawing power.
- Modeling suggests at least 24 in. of unobstructed raised floor. Optimize airflow putting less stress on CRACs. Less than 24 in. will need higher velocities of air.
- Plan for water. It will be used on high-end equipment to reduce energy requirements and hot spots on the raised floor.
- Capture rainwater for on-site storage of water.
- Putting water closer to heat loads will minimize the need for more air conditioning.
- Control hot air rising. How? Ceiling return of hot air, row air curtains, and use cardboard or plastic if you want to really go cheap. Block for recirculation (but make sure you can use sprinklers). Dividers to the ceiling can prevent hot air from escaping into cold aisles.
- Plugged openings (cables, power) not in cold aisle. Block at server or other IT equipment cable tailgates and cable openings in the raised floor to prevent cold air losses and improve efficiencies.
- Two tiles on both hot and cold aisles. Enables both tiles to be pulled up to ease of access to underneath floor. Having tiles with many perforations (holes) in the cold aisle is a must.
- Tile and CRAC placement can be optimized (temperature and airflow) with fluid dynamic analysis. Moving air with less direction changes is best. Watch out for too many perforations, and do not place any perforated tiles within four tiles from CRAC.
- Cables overhead – even power; leave underfloor space to pumping and airflow. When laying out cables, make sure they do not impede airflow if under raised floor (or above). A tray or trough can be made.
- If cables are already under the floor, manage them. Remove cables when no longer needed. Keep "rat's nest" of cables to a minimum.
- Auto lights out. Lights need to be on only when someone is inside data center.
- Shock and vibration support for racks. This is a must do for earthquake regions and can be planned when refreshing equipment or greening the data center.
- Lighting on back of racks for ease of servicing.
- New-generation battery, flywheel, and diesel generator backup. Flywheel can provide a very green way to keep up power system until generators start. Equipment has substantially improved in energy efficiency in last year. Use latest

generation of UPS, flywheel, and generators. Newer generations of UPS are much more efficient than older generations.

- Negotiate with electric utility for going off grid (use generators) in rare peak demand situations. Utilities will pay for the ability to shed load that may potentially cause a brownout or a blackout. The data center can contribute to the electric grid resilience by working with the utility.
- Redundancy design for power and cooling. Eliminate common cause failures wherever possible including UPS by having flywheel and batteries rather than redundant battery UPS.
- Liquid cooling for hot equipment with rear door heat exchanger or side car technologies.
- Enough water cooling taps built into water system for new growth.
- Minimize impedance of piping (water or air) to reduce pumping power required.
- Leak detection under raised floor for water distribution system.
- Variable frequency drives on all pumps and air-conditioning equipment.
- Do not build out all facilities at once. Be modular and strive for high utilization of IT sand facilities equipment.
- Easily displayed power and thermal monitoring for data center. Large displays to highlight success/energy saved. Take pride in using less.
- Where possible, measure overall data-center power unit efficiency, and plan on ways to keep reducing as new state-of-the-art equipment and processes become available.
- Physical: fire protection system FM-200 or wet sprinkler + security system with "man trap" to keep potential intruders from entering the raised floor.
- Liquid side economizer. Efficiently control humidity.
- Thermal storage to optimize use of chillers and reduce energy cost at peak hours.
- High-efficiency pumps, chillers, and fans for cooling towers.
- If possible, use cooling towers in summer, and reuse waste heat to reduce heating energy use in winter.
- When designing a new facility, place the infrastructure in the basement. Position the water piping and electric cables in the ceiling of the basement below the raised floor, leaving the first floor raised flow clear of obstructions.
- Use virtualization for testing and potentially actual disaster recovery. Where possible, architect and build active solutions (production in two places that can nondisruptively add capacity to back up and scale when needed).
- Virtualize servers, storage, and network.
- Link facilities and IT. Offer new types of service-level agreements based on performance/watt and not just performance.
- Be open to cogeneration of electricity. Fuel cells, gas turbines, wind turbines, solar arrays, and even small nuclear reactors (such as the Toshiba 4S) can generate electricity to augment or supply all the needs of a data center.
- Use IT as a green catalyst for organizational efficiencies. Examples are telecommuting, virtualizing desktops, travel optimization, and supply chain management.

- Set both tactical and strategic green goals. Educate team and have the entire team devoted to achieving green goals.
- Celebrate your green successes. Whether it is posting of reductions in electric bill as you walk into the data center or energy efficiency certificates in the annual report, taking pride in accomplishments generates enthusiasm for the next project and fosters an energy-saving culture.

Tools and Information to Help with Green IT

This section lists tools and information available online to help with your analysis of green IT.

DOE DC Pro Tool

The Department of Energy (DOE) Data Center (DC) Pro (Profile) tool is an online software tool designed to help organizations worldwide quickly "diagnose" how energy is being used by their data centers and how they might save energy and money. The tool is available at no cost from https://datacenters.lbl.gov/dcpro.

Carbon Footprint Tool

This site provides a calculator tool for both your personal and business carbon footprint. The web site is http://www.carbonfootprint.com.

ASHRAE Information

The American Society of Heating, Refrigerating and Air-Conditioning Engineers (ASHRAE) has a variety of energy efficiency information available on their web site http://www.ashrae.com.

Server Power Calculator Tools

The various IT vendors provide power calculation tools on their web sites. A general server power calculator tool can be accessed at https://www.libertycenterone.com/blog/quick-server-power-calculator-estimate-server-power-consumption-with-this-browser-based-calculator/.

Computer Manufacturers and Green Computing

All computer manufacturers are busy not only reducing energy use in the servers they produce but also working on making their own data centers green computing examples. There are many case studies for most of the major manufacturers based on news articles. Although there's a marketing aspect to these articles, the emphasis on green IT is significant.

Chip Technology and Green IT

More efficient processors can be a significant energy-saving element, as IBM, Intel, and Advanced Micro Devices all have gotten the green religion. Where chipmakers used to compete entirely on speed, now they also compete on performance per watt. Companies are betting on multicore chip efficiency to fuel interest in new high-end servers. When Intel launched its quad-core Xeon chips beginning in November 2008, it noted that they could deliver 1.8 teraflop peak performance using less than 10,000 W, compared with 800,000 W 10 years ago using Pentium chips. IBM's advanced Power 6™ microprocessor has built-in energy efficiency features such as the ability to reduce power to idle cores, nap mode to power off inactive cores and restore power when needed, and thermal tuning. Almost all modern microprocessors are being designed with "hooks" for virtualization built-in.

How big a difference can more efficient processors make? Princeton University's plasma physics lab, funded by the Department of Energy, cuts 75% of its annual power and cooling bill – from $105,000 in 2003 to $27,000 in 2007 – while improving processing power three to four times. The lower energy use also means it is emitting about 28 fewer tons of carbon dioxide, says Paul Henderson, head of the lab's systems and network group. It did so by replacing a cluster of 200 servers based on AMD Athlon chips with Sun X2100 servers based on dual-core AMD Opteron chips. New generations of IBM servers and supercomputers provide much more performance per watt. Energy efficiency is built into the chip and server designs. High-efficiency power supplies, variable speed fans, and the use of service processors to measure and monitor make new generations of servers very energy efficient compared with older servers. IBM's iDataPlex, designed with energy efficiency in mind for Web 2.0 workloads, uses up to 40% less power than traditional 1U rackable servers.

For every kilowatt of energy consumed by a server, roughly another kilowatt is chewed up to cool it today. Highmark, the largest health insurer in Pennsylvania, uses about 150 blade servers, which reduce the space needed, but their heat and density suck up the cooling. Highmark uses a system that can detect air temperature at the server racks and "tunnel" cooled air to the equipment using special racks from Wright Line, rather than cool the entire room. Highmark takes the environment seriously enough that the toilets in its LEEDs certified eco-designed data center are flushed using rainwater collected from the roof and stored under the parking lot. Highmark is also exploiting large centralized servers such as z10 mainframes to virtualize traditional and Linux servers within a single box sharing resources, thus improving the IT equipment energy efficiency by reducing the number of servers required. Other vendors, such as DegreeControl and, in the summer of 2008, Hewlett-Packard, offered cooling systems that rely on sensors to direct cooling to the needed spot. IBM and other vendors are expanding water cooling solutions as energy efficiency alternatives to traditional air cooling.

Energy Efficiency for Computer Networks

Servers and cooling equipment consume the largest fraction of data-center power by far. Little attention is given to network components, but they also consume power and produce heat. As part of looking at green IT in general, the authors have previously published information on making computer networks green (Lamb, 2009). Use the following energy-saving tips as a way to make your network green:

- Shut off unneeded equipment. If server virtualization has resulted in fewer physical servers, the switch ports that supported the now removed servers are no longer needed. Staff reductions may have resulted in fewer workstations, and therefore fewer switch ports are required. If some switches have a few active ports and others have unused ports, consolidate connections and unplug one of the switches.
- Replace old, inefficient network hardware. Concerns about network energy consumption have led manufacturers to design higher-efficiency power supplies. Newer equipment consumes less power while delivering the same or improved function. Use the local cost of power to evaluate each potential replacement to calculate the payoff period.
- Consolidate multiple small switches, which may have been purchased as the network grew, into a single larger switch. A single high-port-count switch is more energy efficient than many smaller switches.
- Calculate actual power requirements in switches with modular power supplies. Switches may have been over-provisioned when first installed, since power consumption was not a major consideration in the past. Power supplies operate more efficiently at a higher percentage utilization of available capacity. An unneeded supply increases available capacity, so at a given level of utilization, percentage utilization is lower, resulting in reduced efficiency. Put another way, using 40 W of a 50 W supply is much more efficient than using 40 W of a 100 W supply. If possible, remove one or more of the supplies. If the additional supply was put in place to provide redundancy, however, removing it may not be an option.
- Review use of standalone virtual private networks (VPNs), firewalls, and DHCP servers. These standalone appliances have proliferated, and each contains a

power supply, takes up rack space, and produces heat. Moving these functions into a modular switch can reduce power and heat.

- Determine whether 100 Mbps is sufficient for workstation users. Most new workstations come with 1 Gbps Ethernet ports, which consume roughly 2 W more than 100 Mbps. Configuring 1 Gbps on the workstation and on the corresponding switch port adds 4 W to each workstation. While not significant for a small or medium-sized site, the unnecessary power use and heat can add up for a large site.
- Evaluate use of Power over Ethernet (PoE). It is an efficient way to power IP phones, wireless access points, and security cameras. It is not necessary on all switch ports, however, since it cannot be used to power workstations or servers. If PoE is available on all switch ports, make sure that it is configured off for ports that do not use it. When provisioning the power supply of a new modular switch, keep in mind that not all ports will require PoE.
- Properly place all components to efficiently draw in cool air and exhaust hot air. While components may have been placed correctly when first installed, the addition or removal of adjacent equipment may have resulted in less efficient airflow. Use blank panels or move components to fill in gaps in racks that allow cool and hot air to mix.
- Remove unnecessary terminals from switch console ports. Most switch management is done via the network, but a console port terminal may still be in place and powered up even though no longer needed. If for some reason it must be left in place, shut it off.
- Conduct a component-by-component review of your customer's network to identify additional ways to lower energy costs. Many methods of achieving lower energy costs are common sense and will appear obvious once identified. While reviewing the customer's network, you can also suggest equipment upgrades and operational changes that will result in improved network operation.

Appendix B. Comparison of Healthcare Power Generation Methods

It is not the strongest of the species that survives, nor the most intelligent, but the one most responsive to change.

—Charles Darwin

The comparisons in this appendix include renewable methods, e.g., wind turbines, solar, geothermal, etc. Although the topic of establishing green healthcare and green IT should be independent of methods used to generate electricity – with the possible exception of fuel cells which have been proposed as a potential part of green data-center technology – power generation technology is having an impact on the location of data centers. Here's an example: In 2009, Google opened a data center in Council Bluffs, Iowa, close to abundant wind power resources for fulfilling green energy objectives and proximate to fiber-optic communications links. I (John Lamb) grew up in North Dakota, so I proudly emphasize in this appendix that North Dakota is the top wind state (in potential) in the United States. The Great Plains area of the United States has even been referred to as the Saudi Arabia of wind. This overall awareness, sense of responsibility to contribute, and pride in contributing will continue to be very important for our success in driving toward green IT. With high-speed networks (including the Internet) available all over the United States and, for a large part, all over the world, data centers can now be located almost anywhere. And it's not only the location of data networks. High-speed networks have made global resourcing feasible so that computer programmers, analysts, project managers, etc. can now easily be in India or in Chicago. Usually the time zone differences are a bigger impediment to using global resources than any aspect on data communications or network performance.

The significant increase in the use of renewable energy sources (especially solar and wind) over the past 10 years (2007–2017) is a major reason for showing solar and wind energy use for both the 2007 era and the 2017 era. 2007 alternative energy use was detailed in a previous book by the author (Lamb, 2009). This appendix compares the increase from that 2007 era consumption to today's 2017 consump-

© Springer International Publishing AG, part of Springer Nature 2018
N. S. Godbole, J. P. Lamb, *Making Healthcare Green*,
https://doi.org/10.1007/978-3-319-79069-5

tion. The significant increases in solar and wind energy use along with the significant decrease in solar energy cost (worldwide) are highlighted.

The global economic meltdown of 2008 and 2009 has continued to have an impact on the economics of power generation methods. The huge rise in oil prices during the first half of 2008, along with government incentives and advances in solar voltaic and wind power technologies, made those alternative energy sources much more attractive. Then the global economic meltdown drove down oil prices. However, stimulus spending after the Great Recession included large sums for alternative energy as part of the push to create millions of "green" jobs.

As described in this chapter, the cost of generating electricity with solar photovoltaic systems has come down dramatically during the past 10 years. Several of the case studies in Chaps. 10 and 11 describe how hospitals and other healthcare facilities are using solar panels to generate electricity. Healthcare benefits from the use of renewable power generating resources by both reducing pollutants in the air and by reducing the cost of power generation. So, healthcare facilities benefit in several ways when they use green technology.

As indicated in Appendix C, the cost per kWh can vary widely on a worldwide basis. This appendix includes some information on the costs to generate electric power based on power generation method (e.g., coal fired, hydroelectric, gas fired, etc.). This appendix also has information on how estimates on emissions vary based on power generation method. The Appendix C tables give estimates for cost per kWh based on averages for the region.

Cost and Emission Comparisons for Different Power Generation Methods

This information for Tables B.1 and B.2 was compiled by the Pure Energy Systems (PES) group based on 2005 information and given in a previous book by the author (Lamb, 2009). The energy cost comparisons have changed dramatically 12 years later, and Table B.3 shows 2017 energy cost data from the US Department of Energy. The cost of electricity from solar has shown the most significant reduction.

Traditional Power Generation (Lowest Price Listed First)

Table B.1 lists traditional power generation methods along with the 2005 cost range to generate electricity for each method.

Table B.1 Traditional power generation

Method	Cents/ kWh	Limitations and externalities
Gas Currently supplies around 15% of the global electricity demand	3.9–4.4 cents/ kWh	Gas-fired plants are generally quicker and less expensive to build than coal or nuclear, but a relatively high percentage of the cost/kWh is derived from the cost of the fuel. Due to the current (and projected future) upward trend in gas prices, there is uncertainty around the cost/kWh over the lifetime of plants. Gas burns more cleanly than coal, but the gas itself (largely methane) is a potent greenhouse gas. Some energy conversions to calculate your cost of natural gas per kWh. 100 cubic feet (CCF) ~1 Therm = 100,000 BTU ~ 29.3 kWh
Coal Currently supplies around 38% of the global electricity demand	4.8–5.5 cents/ kWh	It is increasingly difficult to build new coal plants in the developed world, due to environmental requirements governing the plants. Growing concern about coal-fired plants in the developing world (China, for instance, imposes less environmental overhead and has large supplies of high sulfur content coal). The supply of coal is plentiful, but the coal generation method is perceived to make a larger contribution to air pollution than the rest of the methods combined
Nuclear Currently supplies around 24% of the global electricity demand	11.1– 14.5 cents/ kWh	Political difficulties in using nuclear in some nations. Risk of widespread (and potentially lethal) contamination upon containment failure. Fuel is plentiful, but problematic. Waste disposal remains a significant problem, and de-commissioning is costly (averaging approximately US $320 million per plant in the United States)

Conventional, Renewable Power Generation (Lowest Price Listed First)

Table B.2 lists conventional, renewable power generation methods along with the 2005 cost range to generate electricity for each method.

Cost Comparison of Energy Sources (2017)

http://www.renewable-energysources.com/

US EIA – US Energy Information Administration (https://www.eia.gov/outlooks/aeo/pdf/electricity_generation.pdf)

Table B.2 Conventional, renewable power generation

Method	Cents/kWh	Limitations and externalities
Wind Currently supplies approximately 1.4% of the global electricity demand. Wind is considered to be about 30% reliable	4.0–6.0 cents/kWh	Wind is currently the only cost-effective alternative energy method but has a number of problems. Wind farms are highly subject to lightning strikes; have high mechanical fatigue failure; are limited in size by hub stress; do not function well, if at all, under conditions of heavy rain, icing conditions, or very cold climates; and are noisy and cannot be insulated for sound reduction due to their size and subsequent loss of wind velocity and power
Geothermal Currently supplies approximately 0.23% of the global electricity demand. Geothermal is considered 90–95% reliable	4.5–30 cents/kWh	New low temperature conversion of heat to electricity is likely to make geothermal substantially more plausible (more shallow drilling possible) and less expensive. Generally, the bigger the plant, the less the cost, and cost also depends upon the depth to be drilled and the temperature at the depth. The higher the temperature, the lower the cost per kWh. Cost may also be affected by where the drilling is to take place as it concerns distance from the grid and another factor may be the permeability of the rock
Hydro Currently supplies around 19.9% of the global electricity demand. Hydro is considered to be 60% reliable	5.1–11.3 cents/kWh	Hydro is currently the only source of renewable energy making substantial contributions to global energy demand. Hydro plants, however, can (obviously) only be built in a limited number of places and can significantly damage aquatic ecosystems
Solar Currently supplies approximately 0.8% of the global electricity demand	15–30 cents/kWh	Solar power has been expensive but soon is expected to drop to as low as 3.5 cents/kWh. Once the silicon shortage is remedied through alternatives to silicon, a solar energy revolution is expected

Table B.3 2017 updates of power generation costs

Power plant type	Cost Cents/kWh
Coal	11–12
Natural gas	5.3–11
Nuclear	9.6
Wind	4.4–20
Solar PV	5.8
Solar thermal	18.4
Geothermal	5.0
Biomass	9.8
Hydro	6.4

Adapted from US DOE (US DOE Annual Energy Outlook 2017)

Worldwide Aspects of Hydroelectricity

Since hydro is currently the source of renewable energy making the most substantial contributions to global energy demand, we'll take a further look at this source of electricity. From Table B.2, hydro supplies around 20% of global electricity demand. Hydro plants, however, can (obviously) only be built in a limited number of places and can significantly damage aquatic ecosystems. Hydropower produces no waste and does not produce carbon dioxide (CO_2), a greenhouse gas. Much of the following information comes from the web site http://en.wikipedia.org/wiki/Hydroelectricity.

Although large hydroelectric installations generate most of the world's hydroelectricity, small hydro schemes are particularly popular in China, which has over 50% of world small hydro capacity. Some jurisdictions throughout the world do not consider large hydro projects to be a sustainable energy source due to human and environmental impacts, though this judgment depends on the definition of sustainability used.

Most hydroelectric power comes from the potential energy of dammed water driving a water turbine and generator. In this case the energy extracted from the water depends on the volume and on the difference in height between the source and the water's outflow. This height difference is called the head. The amount of potential energy in water is proportional to the head. To obtain very high head, water for a hydraulic turbine may be run through a large pipe called a penstock.

Pumped storage hydroelectricity produces electricity to supply high peak demands by moving water between reservoirs at different elevations. At times of low electrical demand, excess generation capacity is used to pump water into the higher reservoir. When there is higher demand, water is released back into the lower reservoir through a turbine. Pumped storage schemes currently provide the only commercially important means of large-scale grid energy storage and improve the daily load factor of the generation system. Hydroelectric plants with no reservoir capacity are called run-of-the-river plants, since it is not then possible to store water. A tidal power plant makes use of the daily rise and fall of water due to tides; such sources are highly predictable and, if conditions permit construction of reservoirs, can also be dispatchable to generate power during high-demand periods.

Annual electric energy production depends on the available water supply. In some installations, the water flow rate can vary by a factor of 10:1 over the course of a year.

Small-Scale Hydroelectric Plants

Small hydro plants are those producing up to 10 MW, although projects up to 30 MW in North America are considered small hydro and have the same regulations. A small hydro plant may be connected to a distribution grid or may provide power only to an isolated community or a single home. Small hydro projects generally do not require the protracted economic, engineering, and environmental studies associated with large projects and often can be completed much more quickly. A small hydro development may be installed along with a project for flood control, irrigation, or other purposes, providing extra revenue for project costs. In areas that formerly used waterwheels for milling and other purposes, often, the site can be redeveloped for electric power production, possibly eliminating the new environmental impact of any demolition operation. Small hydro can be further divided into mini-hydro, units around 1 MW in size, and micro-hydro with units as large as 100 kW down to a couple of kW rating.

Small hydro units ranging from 1 MW to about 30 MW are often available from multiple manufacturers using standardized "water-to-wire" packages; a single contractor can provide all the major mechanical and electrical equipment (turbine, generator, controls, switchgear), selecting from several standard designs to fit the site conditions. Micro-hydro projects use a diverse range of equipment; in the smaller sizes, industrial centrifugal pumps can be used as turbines, with comparatively low purchase cost compared to purpose-built turbines.

Advantages

The major advantage of hydroelectricity is elimination of the cost of fuel. The cost of operating a hydroelectric plant is nearly immune to increases in the cost of fossil fuels such as oil, natural gas, or coal. Fuel is not required and so it need not be imported. Hydroelectric plants tend to have longer economic lives than fuel-fired generation, with some plants now in service having been built 50–100 years ago. Operating labor cost is usually low since plants are automated and have few personnel on site during normal operation.

Where a dam serves multiple purposes, a hydroelectric plant may be added with relatively low construction cost, providing a useful revenue stream to offset the costs of dam operation. It has been calculated that the sale of electricity from China's Three Gorges Dam will cover the construction costs after 5–8 years of full generation.

Greenhouse Gas Emissions

Since hydroelectric dams do not burn fossil fuels, they do not directly produce carbon dioxide (a greenhouse gas). While some carbon dioxide is produced during manufacture and construction of the project, this is a tiny fraction of the operating emissions of equivalent fossil-fuel electricity generation.

Related Activities

Reservoirs created by hydroelectric schemes often provide facilities for water sports and become tourist attractions in themselves. In some countries, farming fish in the reservoirs is common. Multiuse dams installed for irrigation can support the fish farm with relatively constant water supply. Large hydro dams can control floods, which would otherwise affect people living downstream of the project. When dams create large reservoirs and eliminate rapids, boats may be used to improve transportation.

Disadvantages

Recreational users must exercise extreme care when near hydroelectric dams, power plant intakes, and spillways.

Environmental Damage

Hydroelectric projects can be disruptive to surrounding aquatic ecosystems both upstream and downstream of the plant site. For instance, studies have shown that dams along the Atlantic and Pacific coasts of North America have reduced salmon populations by preventing access to spawning grounds upstream, even though most dams in salmon habitat have fish ladders installed. Salmon spawns are also harmed on their migration to sea when they must pass through turbines. This has led to some areas transporting smolt downstream by barge during parts of the year. In some cases dams have been demolished (e.g., the Marmot Dam demolished in 2007, because of impact on fish). Turbine and power plant designs that are easier on aquatic life are an active area of research. Mitigation measures such as fish ladders may be required at new projects or as a condition of re-licensing of existing projects.

Generation of hydroelectric power changes the downstream river environment. Water exiting a turbine usually contains very little suspended sediment, which can lead to scouring of river beds and loss of riverbanks. Since turbine gates are often opened intermittently, rapid or even daily fluctuations in river flow are observed. For example, in the Grand Canyon, the daily cyclic flow variation caused by Glen Canyon Dam was found to be contributing to erosion of sandbars. Dissolved oxygen content of the water may change from pre-construction conditions. Depending on the location, water exiting from turbines is typically much warmer than the pre-dam water, which can change aquatic faunal populations, including endangered species, and prevent natural freezing processes from occurring. Some hydroelectric projects also use canals to divert a river at a shallower gradient to increase the head of the scheme. In some cases, the entire river may be diverted leaving a dry riverbed. Examples include the Tekapo and Pukaki Rivers.

A further concern is the impact of major schemes on birds. Since damming and redirecting the waters of the Platte River in Nebraska for agricultural and energy use, many native and migratory birds such as the piping plover and sandhill crane have become increasingly endangered.

Greenhouse Gas Emission

The reservoirs of power plants in tropical regions may produce substantial amounts of methane and carbon dioxide. This is due to plant material in flooded areas decaying in an anaerobic environment and forming methane, a very potent greenhouse gas. According to the World Commission on Dams report, where the reservoir is large compared to the generating capacity (less than 100 W per square meter of surface area) and no clearing of the forests in the area was undertaken prior to impoundment of the reservoir, greenhouse gas emissions from the reservoir may be higher than those of a conventional oil-fired thermal generation plant. These emissions represent carbon already in the biosphere, not fossil deposits that had been sequestered from the carbon cycle.

In boreal reservoirs of Canada and Northern Europe, however, greenhouse gas emissions are typically only 2–8% of any kind of conventional fossil-fuel thermal generation. A new class of underwater logging operation that targets drowned forests can mitigate the effect of forest decay.

Population Relocation

Another disadvantage of hydroelectric dams is the need to relocate the people living where the reservoirs are planned. In many cases, no amount of compensation can replace ancestral and cultural attachments to places that have spiritual value to the displaced population. Additionally, historically and culturally important sites can be flooded and lost. Such problems have arisen at the Three Gorges Dam project in China, the Clyde Dam in New Zealand, and the Ilısu Dam in Southeastern Turkey.

Comparison with Other Methods of Power Generation

Hydroelectricity eliminates the flue-gas emissions from fossil fuel combustion, including pollutants such as sulfur dioxide, nitric oxide, carbon monoxide, dust, and mercury in coal. Hydroelectricity also avoids the hazards of coal mining and the indirect health effects of coal emissions. Compared to nuclear power, hydroelectricity generates no nuclear waste and has none of the dangers associated with uranium mining nor nuclear leaks. Unlike uranium, hydroelectricity is also a renewable energy source.

Compared to wind farms, hydroelectricity power plants have a more predictable load factor. If the project has a storage reservoir, it can be dispatched to generate power when needed. Hydroelectric plants can be easily regulated to follow variations in power demand.

Unlike fossil-fueled combustion turbines, construction of a hydroelectric plant requires a long lead time for site studies, hydrological studies, and environmental impact assessment. Hydrological data up to 50 years or more is usually required to determine the best sites and operating regimes for a large hydroelectric plant. Unlike plants operated by fuel, such as fossil or nuclear energy, the number of sites that can be economically developed for hydroelectric production is limited; in many areas the most cost-effective sites have already been exploited. New hydro sites tend to be far from population centers and require extensive transmission lines. Hydroelectric generation depends on rainfall in the watershed and may be significantly reduced in years of low rainfall or snowmelt. Long-term energy yield may be affected by climate change. Utilities that primarily use hydroelectric power may spend additional capital to build extra capacity to ensure sufficient power is available in low water years.

In parts of Canada (the provinces of British Columbia, Manitoba, Ontario, Quebec, Newfoundland, and Labrador), hydroelectricity is used so extensively that the word "hydro" is often used to refer to any electricity delivered by a power utility. The government-run power utilities in these provinces are called BC Hydro, Manitoba Hydro, Hydro One (formerly "Ontario Hydro"), Hydro-Québec, and Newfoundland and Labrador Hydro, respectively. Hydro-Québec is the world's largest hydroelectric generating company, with a total installed capacity (2005) of 31,512 MW.

Countries with the Most Hydroelectric Capacity

The ranking of hydroelectric capacity is either by actual annual energy production or by installed capacity power rating. A hydroelectric plant rarely operates at its full power rating over a full year; the ratio between annual average power and installed capacity rating is the load factor. The installed capacity shown in Table B.4 is the sum of all generator nameplate power ratings in 2005 (with 2007 results shown for China). Table B.5 shows the power ratings for the different countries for 2014. Note that China went from 487 TWH annual production in 2007 to 1064 TWH annual production in 2014. Table B.6 gives additional hydroelectric power capacity under construction.

Table B.4 Worldwide annual hydroelectric energy production (2005)

Country	Annual hydroelectric energy production (TWh)	Installed capacity (GW)	Load factor
People's Republic of China (2007)	486.7	145.26	0.37
Canada	350.3	88.974	0.59
Brazil	349.9	69.080	0.56
USA	291.2	79.511	0.42
Russia	157.1	45.000	0.42
Norway	119.8	27.528	0.49
India	112.4	33.600	0.43
Japan	95.0	27.229	0.37
Venezuela	74	–	–
Sweden	61.8	–	–
France	61.5	25.335	0.25

In 2015, hydropower generated 16.6% of the world's total electricity and 70% of all renewable electricity and was expected to increase about 3.1% each year for the next 25 years. Hydropower is produced in 150 countries, with the Asia-Pacific region generating 33% of global hydropower in 2013. China is the largest hydro-electricity producer, with 920 TWH of production in 2013, representing 16.9% of domestic electricity use.

The cost of hydroelectricity is relatively low, making it a competitive source of renewable electricity. The hydro station consumes no water, unlike coal or gas plants. The average cost of electricity from a hydro station larger than 10 MW is 3–5 US cents per kilowatt-hour. With a dam and reservoir, it is also a flexible source of electricity since the amount produced by the station can be changed up or down very quickly to adapt to changing energy demands. Once a hydroelectric complex is constructed, the project produces no direct waste and has a considerably lower output level of greenhouse gases than fossil fuel-powered energy plants.

The ranking of hydroelectric capacity is either by actual annual energy production or by installed capacity power rating. In 2015, hydropower generated 16.6% of the world's total electricity and 70% of all renewable electricity. Hydropower is produced in 150 countries, with the Asia-Pacific region generated 32% of global hydropower in 2010. China is the largest hydroelectricity producer, with 721 TWH of production in 2010, representing around 17% of domestic electricity use. Brazil, Canada, New Zealand, Norway, Paraguay, Austria, Switzerland, and Venezuela have a majority of the internal electric energy production from hydroelectric power. Paraguay produces 100% of its electricity from hydroelectric dams and exports 90% of its production to Brazil and to Argentina. Norway produces 98–99% of its electricity from hydroelectric sources.

A hydroelectric station rarely operates at its full power rating over a full year; the ratio between annual average power and installed capacity rating is the capacity factor. The installed capacity is the sum of all generator nameplate power ratings.

Ten of the Largest Hydroelectric Producers as of 2014

Table B.5 Worldwide annual hydroelectric energy production (2014)

Country	Annual hydroelectric production (TWh)	Installed capacity (GW)	Capacity factor	% of total production
China	1064	311	0.37	18.7%
Canada	383	76	0.59	58.3%
Brazil	373	89	0.56	63.2%
United States	282	102	0.42	6.5%
Russia	177	51	0.42	16.7%
India	132	40	0.43	10.2%
Norway	129	31	0.49	96.0%
Japan	87	50	0.37	8.4%
Venezuela	87	15	0.67	68.3%
France	69	25	0.46	12.2%

Major Projects Under Construction

Table B.6 Major worldwide projects under construction (2015)

Name	Maximum capacity	Country	Construction started	Scheduled completion	Comments
Belo Monte Dam	11,181 MW	Brazil	March 2011	2015	Preliminary construction underway
					Construction suspended 14 days by court order Aug 2012
Siang Upper HE Project	11,000 MW	India	April 2009	2024	Multiphase construction over a period of 15 years. Construction was delayed due to dispute with China
Tasang Dam	7110 MW	Burma	March 2007	2022	Controversial 228 meter tall dam with capacity to produce 35,446 GWh annually
Xiangjiaba Dam	6400 MW	China	November 26, 2006	2015	The last generator was commissioned on July 9, 2014
Grand Ethiopian Renaissance Dam	6000 MW	Ethiopia	2011	2017	Located in the upper Nile Basin, drawing complaint from Egypt
Nuozhadu Dam	5850 MW	China	2006	2017	
Jinping 2 Hydropower Station	4800 MW	China	January 30, 2007	2014	To build this dam, 23 families and 129 local residents need to be moved. It works with Jinping 1 Hydropower Station as a group
Diamer-Bhasha Dam	4500 MW	Pakistan	October 18, 2011	2023	
Jinping 1 Hydropower Station	3600 MW	China	November 11, 2005	2014	The sixth and final generator was commissioned on July 15, 2014
Jirau Power Station	3300 MW	Brazil	2008	2013	Construction halted in March 2011 due to worker riots
Guanyinyan Dam	3000 MW	China	2008	2015	Construction of the roads and spillway started

(continued)

Table B.6 (continued)

Name	Maximum capacity	Country	Construction started	Scheduled completion	Comments
Lianghekou Dam	3000 MW	China	2014	2023	
Dagangshan Dam	2600 MW	China	August 15, 2008	2016	
Liyuan Dam	2400 MW	China	2008	2013	
Tocoma Dam Bolívar State	2160 MW	Venezuela	2004	2014	This power station would be the last development in the Low Caroni Basin, bringing the total to six power stations on the same river, including the 10,000 MW Guri Dam
Ludila Dam	2100 MW	China	2007	2015	Brief construction halt in 2009 for environmental assessment
Shuangjiangkou Dam	2000 MW	China	December 2007	2018	The dam will be 312 m high
Ahai Dam	2000 MW	China	July 27, 2006	2015	
Teles Pires Dam	1820 MW	Brazil	2011	2015	
Site C Dam	1100 MW	Canada	2015	2024	First large dam in western Canada since 1984
Lower Subansiri Dam	2000 MW	India	2007	2016	

Worldwide Aspects of Wind Power

From Table B.2, in 2005 wind power only supplied about 1.5% of global electricity demand. However, since wind is currently the only cost-effective alternative energy method, this section will give an overview of worldwide development of this power source. Much of the following information comes from the web site http://en.wikipedia.org/wiki/Wind_power.

Some companies are looking at wind power as a major source of electric power generation for data centers. For example, in 2009, Google opened one of its first sites in the upper Midwest in Council Bluffs, Iowa, close to abundant wind power resources for fulfilling green energy objectives and proximate to fiber-optic communication links. In general, wind power refers to the conversion of wind energy into a useful form, such as electricity, using wind turbines. At the end of 2007, worldwide capacity of wind-powered generators was 94.1 GW. Although wind produces just a little over 1% of worldwide electricity use, it accounts for approximately 19% of electricity production in Denmark, 9% in Spain and Portugal, and 6% in Germany and the Republic of Ireland (2007 data). Globally, wind power generation increased more than fivefold between 2000 and 2007. Because of the energy crunch that reached critical levels during 2008, wind power is getting a very significant push worldwide, and that push includes television ads and investment by people such as Texas oilman T. Boone Pickens.

Most wind power is generated in the form of electricity. Large-scale wind farms are connected to electrical grids. Individual turbines can provide electricity to isolated locations. In windmills, wind energy is used directly as mechanical energy for pumping water or grinding grain. Wind energy is plentiful, renewable, widely distributed, and clean and reduces greenhouse gas emissions when it displaces fossil-fuel-derived electricity. Therefore, it is considered by experts to be more environmentally friendly than many other energy sources. The intermittency of wind seldom creates problems when using wind power to supply a low proportion of total demand. Where wind is to be used for a moderate fraction of demand, additional costs for compensation of intermittency are considered to be modest.

The multibladed wind turbine atop a lattice tower made of wood or steel was, for many years, a fixture of the landscape throughout rural America. The modern wind turbine was developed beginning in the 1980s, although designs are still under development.

There is an estimated 72 TW of wind energy on the Earth that potentially can be commercially viable. Not all the energy of the wind flowing past a given point can be recovered.

Distribution of Wind Speed and Grid Management

The strength of wind varies, and an average value for a given location does not alone indicate the amount of energy a wind turbine could produce there. To assess the frequency of wind speeds at a particular location, a probability distribution function is often fit to the observed data. Different locations will have different wind speed distributions. The Rayleigh model closely mirrors the actual distribution of hourly wind speeds at many locations.

Because so much power is generated by higher wind speed, much of the energy comes in short bursts. The 2002 Lee Ranch sample is telling; half of the energy available arrived in just 15% of the operating time. The consequence is that wind energy does not have as consistent an output as fuel-fired power plants; utilities that use wind power must provide backup generation for times that the wind is weak. Making wind power more consistent requires that storage technologies must be used to retain the large amount of power generated in the bursts for later use.

Grid Management

Induction generators often used for wind power projects require reactive power for excitation, so substations used in wind power collection systems include substantial capacitor banks for power factor correction. Different types of wind turbine generators behave differently during transmission grid disturbances, so extensive modeling of the dynamic electromechanical characteristics of a new wind farm is required by transmission system operators to ensure predictable stable behavior during system faults. In particular, induction generators cannot support the system voltage during faults, unlike steam or hydro turbine-driven synchronous generators (however properly matched power factor correction capacitors along with electronic control of resonance can support induction generation without grid). Doubly fed machines, or wind turbines with solid-state converters between the turbine generator and the collector system, have generally more desirable properties for grid interconnection. Transmission system operators will supply a wind farm developer with a grid code to specify the requirements for interconnection to the transmission grid. This will include power factor, constancy of frequency, and dynamic behavior of the wind farm turbines during a system fault.

Since wind speed is not constant, a wind farm's annual energy production is never as much as the sum of the generator nameplate ratings multiplied by the total hours in a year. The ratio of actual productivity in a year to this theoretical maximum is called the capacity factor. Typical capacity factors are 20–40%, with values at the upper end of the range in particularly favorable sites. For example, a 1 MW turbine with a capacity factor of 35% will not produce 8760 MWh in a year ($1 \times 24 \times 365$) but only $0.35 \times 24 \times 365 = 3066$ MWh, averaging to 0.35 MW. Online data is available for some locations, and the capacity factor can be calculated from the yearly output.

Unlike fueled generating plants, the capacity factor is limited by the inherent properties of wind. Capacity factors of other types of power plant are based mostly on fuel cost, with a small amount of downtime for maintenance. Nuclear plants have low incremental fuel cost and so are run at full output and achieve a 90% capacity factor. Plants with higher fuel cost are throttled back to follow load. Gas turbine plants using natural gas as fuel may be very expensive to operate and may be run only to meet peak power demand. A gas turbine plant may have an annual capacity factor of 5–25% due to relatively high energy production cost.

According to a 2007 Stanford University study published in the Journal of Applied Meteorology and Climatology, interconnecting ten or more wind farms allows 33–47% of the total energy produced to be used as reliable, baseload electric power, as long as minimum criteria are met for wind speed and turbine height.

Intermittency and Penetration Limits

Because instantaneous electrical generation and consumption must remain in balance to maintain grid stability, this variability can present substantial challenges to incorporating large amounts of wind power into a grid system. Intermittency and the non-dispatchable nature of wind energy production can raise costs for regulation and incremental operating reserve and (at high penetration levels) could require energy demand management, load shedding, or storage solutions. At low levels of wind penetration, fluctuations in load and allowance for failure of large generating units require reserve capacity that can also regulate for variability of wind generation.

Pumped-storage hydroelectricity or other forms of grid energy storage can store energy developed by high-wind periods and release it when needed. Stored energy increases the economic value of wind energy since it can be shifted to displace higher cost generation during peak demand periods. The potential revenue from this arbitrage can offset the cost and losses of storage; the cost of storage may add 25% to the cost of any wind energy stored, but it is not envisaged that this would apply to a large proportion of wind energy generated.

Peak wind speeds may not coincide with peak demand for electrical power. In California and Texas, for example, hot days in summer may have low wind speed and high electrical demand due to air conditioning. In the United Kingdom, however, winter demand is higher than summer demand and so are wind speeds. Solar power tends to be complementary to wind since on most days with no wind there is sun and on most days with no sun there is wind. A demonstration project at the Massachusetts Maritime Academy shows the effect. A combined power plant linking solar, wind, biogas, and hydrostorage is proposed as a way to provide 100% renewable power. The 2006 Energy in Scotland Inquiry report expressed concern that wind power cannot be a sole source of supply and recommends diverse sources of electric energy.

A report from Denmark noted that their wind power network was without power for 54 days during 2002. Wind power advocates argue that these periods of low wind can be dealt with by simply restarting existing power stations that have been held in readiness. The cost of keeping a power station idle is in fact quite low, since the main cost of running a power station is the fuel.

Wind energy "penetration" refers to the fraction of energy produced by wind compared with the total available generation capacity. There is no generally accepted "maximum" level of wind penetration. The limit for a particular grid will depend on the existing generating plants, pricing mechanisms, capacity for storage or demand management, and other factors. An interconnected electricity grid will already include reserve generating and transmission capacity to allow for equipment failures; this reserve capacity can also serve to regulate for the varying power generation by wind plants. Studies have indicated that 20% of the total electrical energy consumption may be incorporated with minimal difficulty. These studies have been for locations with geographically dispersed wind farms, some degree of dispatchable energy, or hydropower with storage capacity, demand management, and interconnection to a large grid area export of electricity when needed. Beyond this level, there are few technical limits, but the economic implications become more significant.

At present, few grid systems have penetration of wind energy above 5%: Denmark (values over 18%), Spain and Portugal (values over 9%), and Germany and the Republic of Ireland (values over 6%). The Danish grid is heavily interconnected to the European electrical grid, and it has solved grid management problems by exporting almost half of its wind power to Norway. The correlation between electricity export and wind power production is very strong.

Denmark has active plans to increase the percentage of power generated to over 50%. A study commissioned by the state of Minnesota considered penetration of up to 25% and concluded that integration issues would be manageable and have incremental costs of less than one-half cent ($0.0045) per kWh.

ESB National Grid, Ireland's electric utility, in a 2004 study, concluded that to meet the renewable energy targets set by the EU in 2001 would "increase electricity generation costs by a modest 15%."

Good selection of a wind turbine site is critical to economic development of wind power. Aside from the availability of wind itself, other factors include the availability of transmission lines, value of energy to be produced, cost of land acquisition, land use considerations, and environmental impact of construction and operations. Offshore locations may offset their higher construction cost with higher annual load factors, thereby reducing cost of energy produced. Wind farm designers use specialized wind energy software applications to evaluate the impact of these issues on a given wind farm design.

Offshore Windfarms

On December 21, 2007, Q7, a 120 MW offshore wind farm with a construction budget of €383 million, exported first power to the Dutch grid, which was a milestone for the offshore wind industry. Q7 was the first offshore wind farm to be financed by a nonrecourse loan (project finance). The project comprised of 60 2 MW V80 Vestas machines and features monopile foundation to a depth of between 18 and 23 m at a distance of about 23 km off the Dutch coast.

Utilization of Wind Power

There are now many thousands of wind turbines operating, with a total capacity of 73,904 MW of which wind power in Europe accounts for 65% (2006). Wind power was the fastest-growing energy source at the end of 2004. World wind generation capacity more than quadrupled between 2000 and 2006. Eighty-one percent of wind power installations are in the United States and Europe, but the share of the top five countries in terms of new installations fell from 71% in 2004 to 62% in 2006.

Countries with the Most Installed Wind Power Capacity

In 2016, the countries with the highest total installed capacity were China, the United States, Germany, India, Spain, the United Kingdom, France, Canada, Brazil, and Italy. Table B.7 lists the rankings.

As of the end of 2016, the worldwide total cumulative installed electricity generation capacity from wind power amounted to 486,790 MW, an increase of 12.5% compared to the previous year. Installations increased by 54,642 MW, 63,330 MW, 51,675 MW, and 36,023 MW in 2016, 2015, 2014, and 2013, respectively.

Since 2010, more than half of all new wind power was added outside of the traditional markets of Europe and North America, mainly driven by the continuing boom in China and India. At the end of 2015, China had 145 GW of wind power installed. In 2015, China installed close to half of the world's added wind power capacity.

Several countries have achieved relatively high levels of wind power penetration, such as 39% of stationary electricity production in Denmark, 18% in Portugal, 16% in Spain, 14% in Ireland, and 9% in Germany in 2010. As of 2011, 83 countries around the world are using wind power on a commercial basis. Wind power's share of worldwide electricity usage at the end of 2014 was 3.1%.

Installed Wind Power Capacity

This section provides end-of-year figures of worldwide installed wind power capacity by country. The data is sourced from Global Wind Energy Council. In 2015, global wind power capacity increased by 63,330 MW or 17.14% from 369,553 to 432,883 MW.

As seen in Table B.7, the worldwide wind power capacity was 432.4 GW at the end of 2015, up from 74.1 GW at the end of 2006, giving a net growth rate of more than 20% per year.

Denmark generates nearly one-fifth of its electricity with wind turbines – the highest percentage of any country – and is fifth in the world in total wind power generation. Denmark is prominent in the manufacturing and use of wind turbines, with a commitment made in the 1970s to eventually produce half of the country's power by wind.

In recent years, the United States has added more wind energy to its grid than any other country; US wind power capacity grew by 45% to 16.8 GW in 2007. Texas has become the largest wind energy-producing state, surpassing California. In 2007, the state expects to add 2 GW to its existing capacity of approximately 4.5 GW. Iowa and Minnesota are expected to each produce 1 GW by late 2007. Wind power generation in the United States was up 31.8% in February 2007 from February 2006. The average output of 1 MW of wind power is equivalent to the average electricity consumption of about 250 American households. According to the American Wind Energy Association, wind generated enough electricity in 2008 to power just over 1% (4.5 million households) of total electricity in the United States, up from less than 0.1% in 1999. The US Department of Energy studies have concluded wind harvested in Texas, Kansas, and North Dakota could provide enough electricity to power the entire nation and that offshore wind farms could do the same job. Since the author grew in North Dakota (50th of the 50 states in tourist attraction), this is an opportunity to put in a plug for my home state since it is number one in wind energy potential.

Table B.7 Worldwide installed wind power capacity (2006–2016)

Installed wind power capacity (MW)

No.	Nation	2006	2007	2008	2009	2010	2011	2012	2013	2014	2015	2016
1	China	2599	5912	12,210	25,104	44,733	62,733	75,564	91,412	114,763	145,104	168,690
–	European Union	48,122	56,614	65,255	74,919	84,278	93,957	106,454	117,384	128,752	141,579	153,730
2	United States	11,603	16,819	25,170	35,159	40,200	46,919	60,007	61,110	65,879	74,472	82,183
3	Germany	20,622	22,247	23,903	25,777	27,214	29,060	31,332	34,250	39,165	44,947	50,019
4	India	6270	7850	9587	10,925	13,064	16,084	18,421	20,150	22,465	27,151	28,665
5	Spain	11,630	15,145	16,740	19,149	20,676	21,674	22,796	22,959	22,987	23,025	23,075
6	United Kingdom	1963	2389	3288	4070	5203	6540	8445	10,711	12,440	13,603	14,542
7	France	1589	2477	3426	4410	5660	6800	7196	8243	9285	10,358	12,065
8	Canada	1460	1846	2369	3319	4008	5265	6200	7823	9694	11,205	11,898
9	Brazil	237	247	339	606	932	1509	2508	3466	5939	8715	10,740
10	Italy	2123	2726	3537	4850	5797	6747	8144	8558	8663	8958	9257
11	Sweden	571	831	1067	1560	2163	2970	3745	4382	5425	6025	6519
12	Turkey	65	207	433	801	1329	1799	2312	2958	3763	4718	6081
13	Poland	153	276	472	725	1107	1616	2497	3390	3834	5100	5782
14	Portugal	1716	2130	2862	3535	3702	4083	4525	4730	4914	5079	5316
15	Denmark	3140	3129	3164	3465	3752	3871	4162	4807	4845	5063	5227
16	Netherlands	1571	1759	2237	2223	2237	2328	2391	2671	2805	3431	4328
17	Australia	651	824	1306	1712	1991	2176	2584	3239	3806	4187	4327
18	Mexico	84	85	85	520	733	873	1370	1859	2551	3073	3527
19	Japan	1309	1528	1880	2056	2304	2501	2614	2669	2789	3038	3234
20	Romania	2	7	10	14.1	462	982	1905	2599	2953	2976	3028
21	Ireland	746	805	1245	1260	1379	1614	1738	2049	2272	2486	2830
22	Austria	965	982	995	995	1011	1084	1378	1684	2095	2412	2632
23	Belgium	194	287	384	563	911	1078	1375	1651	1959	2229	2386
24	Greece	758	873	990	1087	1208	1629	1749	1866	1980	2152	2374

Installed wind power capacity (MW)

No.	Nation	2006	2007	2008	2009	2010	2011	2012	2013	2014	2015	2016
25	Finland	86	110	143	147	197	199	288	447	627	1005	1539
26	South Africa	–	–	–	–	–	–	–	10	570	1053	1471
27	Chile	–	–	–	20	168	172	205	331	836	933	1424
28	Uruguay	–	–	–	–	–	43	56	59	701	845	1210
29	South Korea	176	192	278	348	379	407	483	561	609	835	1036
30	Morocco	64	125	125	253	286	291	291	487	787	787	882
31	Norway	325	333	428	431	441	512	704	811	819	838	838
32	Bulgaria	36	70	120	177	500	612	674	681	691	691	691
33	Taiwan	188	280	358	436	519	564	564	614	633	647	682
34	New Zealand	171	322	325	497	530	623	623	623	623	623	623
35	Egypt	230	310	390	430	550	550	550	550	610	610	610
36	Pakistan	–	–	–	–	–	–	–	–	–	255	591
37	Ukraine	86	89	90	94	87	151	302	371	498	514	526
38	Lithuania	56	50	54	91	163	203	263	279	279	315	493
39	Croatia	n/a	n/a	69.4	104	152	187.4	207.1	302	347	387	422
40	Hungary	61	65	127	201	295	329	329	329	329	329	329
41	Ethiopia	–	–	–	–	–	23	81	171	171	324	324
42	Estonia	31.8	59	78	142	149	184	269	280	302	303	310
43	Costa Rica	–	–	74	123	119	132	147	148	198	268	298
44	Czech Republic	57	116	150	192	215	217	260	269	281	281	281
45	Argentina	–	–	–	–	–	113	167	218	271	279	279
46	Panama	–	–	–	–	–	–	–	–	35	270	270
47	Tunisia	–	–	–	–	–	54	104	245	245	245	245
48	Peru	–	–	–	–	–	–	–	146	148	148	241

(continued)

Table B.7 (continued)

Installed wind power capacity (MW)

No.	Nation	2006	2007	2008	2009	2010	2011	2012	2013	2014	2015	2016
49	Thailand	–	–	–	–	–	7	112	223	223	223	223
50	Philippines	–	–	–	–	–	–	–	66	216	216	216
51	Nicaragua	–	–	–	–	–	62	102	146	186	186	186
52	Jordan	–	–	–	–	–	–	–	–	2	119	185
53	Honduras	–	–	–	–	–	–	102	102	126	176	176
54	Cyprus	–	–	–	–	–	–	–	–	147	158	158
55	Iran	47	67	82	91	91	91	91	91	n.a.	n.a.	134
56	Sri Lanka	–	–	–	–	–	–	63	63	62	62	
57	Latvia	–	–	–	–	–	–	–	–	58	58	
58	Luxembourg	–	–	–	–	–	–	–	50	n.a.	n.a.	
59	Mongolia	–	–	–	–	–	–	30	–	n.a.	n.a.	
60	Venezuela	–	–	–	–	–	24	24	24	24	n.a.	
61	Cape Verde	–	–	–	–	–	–	191	250	250	250	
–	Caribbean	–	–	–	–	–	–	12	12	12	12	
–	Pacific Islands	–	–	–	–	–	12	12	12	12	12	
–	Rest of Europe	–	–	–	–	–	–	3815	4956	5715	6543	
–	Rest of Latin America and Caribbean	–	–	–	–	–	54	54	250	250	285	
–	Rest of Africa and Middle East	–	–	–	–	–	–	1165	1255	129	151	
–	Rest of Asia	–	–	–	–	–	–	71	87	–	167	
	World total capacity (MW)	74,151	93,927	121,188	157,899	197,637	238,035	282,482	318,596	369,553	432,419	

Table B.8 Top wind states and annual wind energy potential in billions of kilowatt hours

Rank	State	Annual potential (Billion kWh)
1	North Dakota	1210
2	Texas	1190
3	Kansas	1070
4	South Dakota	1030
5	Montana	1020
6	Nebraska	868
7	Wyoming	747
8	Oklahoma	725
9	Minnesota	657
10	Iowa	551
11	Colorado	481
12	New Mexico	435
13	Idaho	73
14	Michigan	65
15	New York	62
16	Illinois	61
17	California	59
18	Wisconsin	58
19	Maine	56
20	Missouri	52

Top Wind States (North Dakota #1)

With the push started in 2008 by Texas oilman, T. Boone Pickens, to cut the United States' demand for foreign oil, wind energy has been significantly promoted in television ads. As stated above, Texas, Kansas, and North Dakota alone could provide via wind power all the electricity needed to power the entire United States. Table B.8 gives the top 20 states in wind power potential. North Dakota (the co-author's home state) is listed as number 1, followed by Texas and Kansas.

India ranks fourth in the world with a total wind power capacity of 8000 MW in 2007, or 3% of all electricity produced in India. The World Wind Energy Conference in New Delhi in November 2006 has given additional impetus to the Indian wind industry. The wind farm near Muppandal, Tamil Nadu, India, provides an impoverished village with energy.

In 2005, China announced it would build a 1000 MW wind farm in Hebei for completion in 2020. China reportedly has set a generating target of 20,000 MW by 2020 from renewable energy sources – it says indigenous wind power could generate up to 253,000 MW. Following the World Wind Energy Conference in November 2004, organized by the Chinese and the World Wind Energy Association, a Chinese renewable energy law was adopted. In late 2005, the Chinese government increased the official wind energy target for the year 2020 from 20 to 30 GW.

Mexico recently opened La Venta II wind power project as an important step in reducing Mexico's consumption of fossil fuels. The 88 MW project is the first of its kind in Mexico and will provide 13% of the electricity needs of the state of Oaxaca. By 2012, the project had a capacity of 3500 MW.

Another growing market is Brazil, with a wind potential of 143 GW. The federal government has created an incentive program, called Proinfa, and built a production capacity of 3300 MW of renewable energy in 2008, of which 1422 MW was through wind energy. The program seeks to produce 10% of Brazilian electricity through renewable sources.

South Africa has a proposed station situated on the West Coast north of the Olifants River mouth near the town of Koekenaap, east of Vredendal in the Western Cape province. The station is proposed to have a total output of 100 MW although there are negotiations to double this capacity. The plant became operational in 2010.

Canada experienced rapid growth of wind capacity between 2000 and 2006, with total installed capacity increasing from 137 to 1451 MW and showing an annual growth rate of 38%. Particularly rapid growth was seen in 2006, with total capacity doubling from the 684 MW at the end of 2005. This growth was fed by measures including installation targets, economic incentives, and political support. For example, the Ontario government announced that it will introduce a feed-in tariff for wind power, referred to as "standard offer contracts," which may boost the wind industry across the province. In Quebec, the provincially owned electric utility purchased an additional 2000 MW in 2013.

Small-Scale Wind Power

Small wind generation systems with capacities of 100 kW or less are usually used to power homes, farms, and small businesses. Isolated communities that otherwise rely on diesel generators may use wind turbines to displace diesel fuel consumption. Individuals purchase these systems to reduce or eliminate their electricity bills or simply to generate their own clean power.

Wind turbines have been used for household electricity generation in conjunction with battery storage over many decades in remote areas. Increasingly, US consumers are choosing to purchase grid-connected turbines in the 1–10 KW range to power their whole homes. Household generator units of more than 1 kW are now functioning in several countries and in every state in the United States.

Grid-connected wind turbines may use grid energy storage, displacing purchased energy with local production when available. Off-grid system users either adapt to intermittent power or use batteries and photovoltaic or diesel systems to supplement the wind turbine.

In urban locations, where it is difficult to obtain predictable or large amounts of wind energy, smaller systems may still be used to run low power equipment. Equipment such as parking meters or wireless internet gateways may be powered by a wind turbine that charges a small battery, replacing the need for a connection to the power grid.

References: see the large list of references at http://en.wikipedia.org/wiki/ Wind_power

Worldwide Aspects of Solar Power

From Table B.2, solar power supplies less than 1% of global electricity demand. The term solar energy refers to the utilization of the radiant energy from the sun. Solar power is used interchangeably with solar energy but refers more specifically to the conversion of sunlight into electricity, either by photovoltaics and concentrating solar thermal devices or by one of several experimental technologies such as thermoelectric converters, solar chimneys, or solar ponds. Since solar power, along with wind power, is the alternative power generation method currently receiving most attention, this section will give an overview of worldwide development of this power source. Much of the following information comes from the web site http:// en.wikipedia.org/wiki/Solar_energy.

The Earth receives 174 petawatts (PW) of incoming solar radiation (insolation) at the upper atmosphere. Approximately 30% is reflected back to space, while the rest is absorbed by clouds, oceans, and land masses. The spectrum of solar light at the Earth's surface is mostly spread across the visible and near-infrared ranges with a small part in the near-ultraviolet.

Solar radiation along with secondary solar resources such as wind and wave power, hydroelectricity, and biomass account for over 99.9% of the available flow of renewable energy on Earth. The total solar energy absorbed by the Earth's atmosphere, oceans, and land masses is approximately 3850 zettajoules (ZJ) per year. In 2002, this was more energy in 1 h than the world used in 1 year. Photosynthesis captures approximately 3 ZJ per year in biomass. The amount of solar energy reaching the surface of the planet is so vast that in 1 year it is about twice as much as will ever be obtained from all of the Earth's nonrenewable resources of coal, oil, natural gas, and mined uranium combined.

Applications of Solar Energy Technology

Technologies that use secondary solar resources such as biomass, wind, waves, and ocean thermal gradients can be included in a broader description of solar energy, but only primary resource applications are discussed here. Because the performance of solar technologies varies widely between regions, they should be deployed in a way that carefully considers these variations.

Solar technologies are broadly characterized as either passive or active depending on the way they capture, convert, and distribute sunlight. Active solar techniques use photovoltaic panels, pumps, and fans to convert sunlight into useful outputs. Passive solar techniques include selecting materials with favorable thermal properties, designing spaces that naturally circulate air, and referencing the position of a building to the sun. Active solar technologies increase the supply of energy and are considered supply side technologies, while passive solar technologies reduce the need for alternate resources and are generally considered demand-side technologies.

Solar Electricity

Sunlight can be converted into electricity using photovoltaics (PV), concentrating solar power (CSP), and various experimental technologies. PV has mainly been used to power small- and medium-sized applications, from the calculator powered by a single solar cell to off-grid homes powered by a photovoltaic array. For large-scale generation, CSP plants like SEGS have been the norm, but recently multi-megawatt PV plants are becoming common. Completed in 2007, the 14 MW power station in Clark County, Nevada, and the 20 MW site in Beneixama, Spain, are characteristic of the trend toward larger photovoltaic power stations in the United States and Europe.

Photovoltaics

A solar cell, or photovoltaic cell (PV), is a device that converts light into direct current using the photoelectric effect. The first solar cell was constructed by Charles Fritts in the 1880s. Although the prototype selenium cells converted less than 1% of incident light into electricity, both Ernst Werner von Siemens and James Clerk Maxwell recognized the importance of this discovery. Following the work of Russell Ohl in the 1940s, researchers Gerald Pearson, Calvin Fuller, and Daryl Chapin created the silicon solar cell in 1954. These early solar cells cost 286 USD/watt and reached efficiencies of 4.5–6%.

The earliest significant application of solar cells was as a backup power source to the Vanguard I satellite, which allowed it to continue transmitting for over a year after its chemical battery was exhausted. The successful operation of solar cells on this mission was duplicated in many other Soviet and American satellites, and by the late 1960s, PV had become the established source of power for them. Photovoltaics went on to play an essential part in the success of early commercial satellites such as Telstar, and they remain vital to the telecommunications infrastructure today.

The high cost of solar cells limited terrestrial uses throughout the 1960s. This changed in the early 1970s when prices reached levels that made PV generation competitive in remote areas without grid access. Early terrestrial uses included powering telecommunication stations, offshore oil rigs, navigational buoys, and railroad crossings. These off-grid applications have proven very successful and accounted for over half of worldwide installed capacity until 2004.

Building-integrated photovoltaics cover the roofs of the increasing number of homes. The 1973 oil crisis stimulated a rapid rise in the production of PV during the

1970s and early 1980s. Economies of scale which resulted from increasing production along with improvements in system performance brought the price of PV down from 100 USD/watt in 1971 to 7 USD/watt in 1985. Steadily falling oil prices during the early 1980s led to a reduction in funding for photovoltaic R&D and a discontinuation of the tax credits associated with the Energy Tax Act of 1978. These factors moderated growth to approximately 15% per year from 1984 to 1996.

Since the mid-1990s, leadership in the PV sector has shifted from the United States to Japan and Germany. Between 1992 and 1994, Japan increased R&D funding, established net metering guidelines, and introduced a subsidy program to encourage the installation of residential PV systems. As a result, PV installations in the country climbed from 31.2 MW in 1994 to 318 MW in 1999, and worldwide production growth increased to 30% in the late 1990s.

Germany has become the leading PV market worldwide since revising its feed-in tariff system as part of the Renewable Energy Sources Act. Installed PV capacity has risen from 100 MW in 2000 to approximately 4150 MW at the end of 2007. Spain has become the third largest PV market after adopting a similar feed-in tariff structure in 2004, while France, Italy, South Korea, and the United States have seen rapid growth recently due to various incentive programs and local market conditions. World solar photovoltaic (PV) market installations reached a record high of 2.8 GWp in 2007.

The three leading countries (Germany, Japan, and the United States) represent nearly 89% of the total worldwide PV installed capacity. On August 1, 2007, a Wednesday, word was published of construction of a production facility in China, which is projected to be one of the largest wafer factories in the world, with a peak capacity of around 1500 MW. Germany was the fastest-growing major PV market in the world during 2006 and 2007. In 2007, over 1.3 GWp of PV was installed. The German PV industry generates over 10,000 jobs in production, distribution, and installation. By the end of 2006, nearly 88% of all solar PV installations in the EU were in grid-tied applications in Germany. The balance is off-grid (or standalone) systems. Photovoltaic power capacity is measured as maximum power output under standardized test conditions (STC) in "Wp" (watts peak). The actual power output at a particular point in time may be less than or greater than this standardized, or "rated," value, depending on geographical location, time of day, weather conditions, and other factors. Solar photovoltaic array capacity factors are typically under 25%, which is lower than many other industrial sources of electricity. Therefore the 2006 installed base peak output would have provided an average output of 1.2 GW (assuming $20\% \times 5862$ MWp). This represented 0.06% of global demand at the time.

Countries with the Most Installed Photovoltaic Peak Power Capacity

In 2007, the countries with the highest photovoltaic peak power capacity were Germany, Japan, the United States, Spain, and Italy. Table B.9 lists the rankings based on photovoltaic peak power capacity estimated at the end of 2007. Table B.10

Table B.9 Worldwide photovoltaic peak power capacity (end of 2007)

Country/ region	Installed off grid in 2007	Installed on grid in 2007	Total installed in 2007	Total off grid	Total on grid	Total end of 2007	Total Wp/ capita total
World	127.86	2129.778	2257.638	662.34	7178.392	7840.732	
Germany	35	1100	1135	35	3827	3862	46.8
Japan	1.562	208.8	210.395	90.15	1828.744	1918.894	15
United States	55	151.5	206.5	325	505.5	830.5	2.8
Spain	22	490	512	29.8	625.2	655	15.1
Italy	0.3	69.9	70.2	13.1	107.1	120.2	2.1
Australia	5.91	6.28	12.19	66.446	16.045	82.491	4.1
South Korea	0	42.868	42.868	5.943	71.658	77.601	1.6
France	0.993	30.306	31.299	22.547	52.685	75.232	1.2
Netherlands	0.582	1.023	1.605	5.3	48	53.3	3.3
Switzerland	0.2	6.3	6.5	3.6	32.6	36.2	4.9
Austria	0.055	2.061	2.116	3.224	24.477	27.701	3.4
Canada	3.888	1.403	5.291	22.864	2.911	25.775	0.8
Mexico	0.869	0.15	1.019	20.45	0.3	20.75	0.2
United Kingdom	0.16	3.65	3.81	1.47	16.62	18.09	0.3
Portugal	0.2	14.254	14.454	2.841	15.029	17.87	1.7
Norway	0.32	0.004	0.324	7.86	0.132	7.992	1.7
Sweden	0.271	1.121	1.392	4.566	1.676	6.242	0.7
Denmark	0.05	0.125	0.175	0.385	2.69	3.075	0.6
Israel	0.5	0	0.5	1.794	0.025	1.819	0.3

lists the rankings for 2016. This information is from http://en.wikipedia.org/wiki/Photovoltaics. Note that from Table B.9, the Wp/capita (watts peak per capita) is far higher for Germany than for any other country.

Many industrialized nations have installed significant solar power capacity into their electrical grids to supplement or provide an alternative to conventional energy sources, while an increasing number of less developed nations have turned to solar to reduce dependence on expensive imported fuels. Long-distance transmission allows remote renewable energy resources to displace fossil fuel consumption. Solar power plants use one of two technologies:

Photovoltaic (PV) systems use solar panels, either on rooftops or in ground-mounted solar farms, converting sunlight directly into electric power. Concentrated solar power (CSP, also known as "concentrated solar thermal") plants use solar thermal energy to make steam that is thereafter converted into electricity by a turbine.

Worldwide growth of photovoltaics is extremely dynamic and varies strongly by country. By the end of 2016, cumulative photovoltaic capacity increased by more than 75 gigawatts (GW) and reached at least 303 GW, sufficient to supply 1.8% of the world's total electricity consumption. The top installers of 2016 were China, the United States, and India. There are more than 24 countries around the world with a cumulative PV capacity of more than one gigawatt. Austria, Chile, and South Africa all crossed the one-gigawatt mark in 2016. The available solar PV capacity in Honduras is now sufficient to supply 12.5% of the nation's electrical power, while Italy, Germany, and Greece can produce between 7% and 8% of their respective domestic electricity consumption.

After an almost two-decade-long hiatus, deployment of CSP resumed in 2007, with significant growth only in the most recent years. However, the design for several new projects is being changed to cheaper photovoltaics. Most operational CSP stations are located in Spain and the United States, while large solar farms using photovoltaics are being constructed in an expanding list of geographic regions.

As of January 2017, the largest solar power plants in the world are the 850 MW Longyangxia Dam Solar Park in China for PV and the 377 MW Ivanpah Solar Power Facility in the United States for CSP. Other large CSP facilities include the 354 megawatt (MW) Solar Energy Generating Systems power installation in the United States, Solnova Solar Power Station (Spain, 150 MW), Andasol Solar Power Station (Spain, 150 MW), and the first part of Shams Solar Power Station (United Arab Emirates, 100 MW).

Installed PV Solar Power in 2016

Table B.10 Worldwide photovoltaic peak power capacity (end of 2016)

#	Nation	Total capacity (GW)	Added capacity (GW)	% of total consumption
1	China	78.07	34.54	1.07%
2	Japan	42.75	8.6	4.9% (2016)
3	Germany	41.22	1.52	6.2–6.7%
4	United States	40.3	14.73	1.4%
5	Italy	19.28	0.37	7.5% (2014)
6	United Kingdom	11.63	1.97	3.4%
7	India	9.01	3.97	
8	France	7.13	0.56	1.6%
9	Australia	5.9	0.84	2.4% (2015)
10	Spain	5.49	0.06	3.2% (2013)
11	South Korea	4.35	0.85	
12	Belgium	3.42	0.17	
13	Canada	2.72	0.2	
14	Greece	2.6	0	
15	Thailand	2.15	0.73	
16	Czech Republic	2.1	0	
17	Netherlands	2.1	0.53	
18	Switzerland	1.64	0.25	

(continued)

Table B.10 (continued)

#	Nation	Total capacity (GW)	Added capacity (GW)	% of total consumption
19	Chile	1.61	0.75	
20	South Africa	1.45	0.54	
21	Taiwan	1.38	0.37	
22	Romania	1.3	0	
23	Austria	1.08	0.15	
24	Bulgaria	1.0	0	
25	Pakistan[a]	1.0	0.6	
26	Israel	0.91	0.13	
27	Philippines	0.9	0.76	
28	Denmark	0.9	0.07	2.8% (2016)[a]
29	Turkey	0.83	0.58	
30	Portugal	0.51	0.06	
31	Honduras[a]	0.39	0.39	
32	Algeria	0.35	0.05	
33	Mexico	0.32	0.15	
34	Malaysia	0.29	0.05	
35	Sweden	0.18	0.06	
36	Norway	0.027	0.011	
37	Finland	0.015	0.010	
	World total PV capacity	303	75	1.8%

[a]Indicates 2015 data

Concentrating Solar Power

Solar troughs are the most widely deployed and the most cost-effective CSP technology. Concentrated sunlight has been used to perform useful tasks since the time of ancient China. A legend claims that Archimedes used polished shields to concentrate sunlight on the invading Roman fleet and repel them from Syracuse. Auguste Mouchout used a parabolic trough to produce steam for the first solar steam engine in 1866, and subsequent developments led to the use of concentrating solar-powered devices for irrigation, refrigeration, and locomotion.

Concentrating solar power (CSP) systems use lenses or mirrors and tracking systems to focus a large area of sunlight into a small beam. The concentrated light is then used as a heat source for a conventional power plant. A wide range of concentrating technologies exist; the most developed are the solar trough, parabolic dish, and solar power tower. These methods vary in the way they track the sun and focus light. In all these systems, a working fluid is heated by the concentrated sunlight and is then used for power generation or energy storage.

The PS10 concentrates sunlight from a field of heliostats on a central tower. A solar trough consists of a linear parabolic reflector that concentrates light onto a receiver positioned along the reflector's focal line. The reflector is made to follow the sun during the daylight hours by tracking along a single axis. Trough systems provide the best land-use factor of any solar technology. The SEGS plants in California and Acciona's Nevada Solar One near Boulder City, Nevada, are representatives of this technology.

A parabolic dish system consists of a stand-alone parabolic reflector that concentrates light onto a receiver positioned at the reflector's focal point. The reflector tracks the sun along two axes. Parabolic dish systems give the highest efficiency among CSP technologies. The 50 kW Big Dish in Canberra, Australia, is an example of this technology.

A solar power tower uses an array of tracking reflectors (heliostats) to concentrate light on a central receiver atop a tower. Power towers are less advanced than trough systems but offer higher efficiency and better energy storage capability. The Solar Two in Barstow, California, and the Planta Solar 10 in Sanlucar la Mayor, Spain, are representatives of this technology.

Experimental Solar Power and Storage Methods

A solar updraft tower (also known as a solar chimney or solar tower) consists of a large greenhouse that funnels into a central tower. As sunlight shines on the greenhouse, the air inside is heated and expands. The expanding air flows toward the central tower, where a turbine converts the airflow into electricity. A 50 kW prototype was constructed in Ciudad Real, Spain, and operated for 8 years before decommissioning in 1989.

Space solar power systems would use a large solar array in geosynchronous orbit to collect sunlight and beam this energy in the form of microwave radiation to receivers (rectennas) on Earth for distribution. This concept was first proposed by Dr. Peter Glaser in 1968, and since then a wide variety of systems have been studied with both photovoltaic and concentrating solar thermal technologies being proposed. Although still in the concept stage, these systems offer the possibility of delivering power approximately 96% of the time.

Energy Storage Methods

Solar Two's thermal storage system generated electricity during cloudy weather and at night. Storage is an important issue in the development of solar energy because modern energy systems usually assume continuous availability of energy. Solar energy is not available at night, and the performance of solar power systems is affected by unpredictable weather patterns; therefore, storage media or backup power systems must be used.

Thermal mass systems can store solar energy in the form of heat at domestically useful temperatures for daily or seasonal durations. Thermal storage systems generally use readily available materials with high specific heat capacities such as water, earth, and stone. Well-designed systems can lower peak demand, shift time-of-use to off-peak hours, and reduce overall heating and cooling requirements.

Phase change materials such as paraffin wax and Glauber's salt are another thermal storage media. These materials are inexpensive and readily available and can deliver domestically useful temperatures (approximately 64 °C). The "Dover House" (in Dover, Massachusetts) was the first to use a Glauber's salt heating system, in 1948.

Solar energy can be stored at high temperatures using molten salts. Salts are an effective storage medium because they are low cost, have a high specific heat capacity, and can deliver heat at temperatures compatible with conventional power systems. The Solar Two used this method of energy storage, allowing it to store 1.44 TJ in its 68 m^3 storage tank with an annual storage efficiency of about 99%.

Off-grid PV systems have traditionally used rechargeable batteries to store excess electricity. With grid-tied systems, excess electricity can be sent to the transmission grid. Net metering programs give these systems a credit for the electricity they deliver to the grid. This credit offsets electricity provided from the grid when the system cannot meet demand, effectively using the grid as a storage mechanism.

Pumped-storage hydroelectricity stores energy in the form of water pumped when energy is available from a lower elevation reservoir to a higher elevation one. The energy is recovered when demand is high by releasing the water to run through a hydroelectric power generator. See the large list of references at http://en.wikipedia.org/wiki/Solar_energy.

The Future for Solar Power

The sun is the force behind all of the Earth's energy. Photosynthesis captured the energy that is burned in fossil fuels. The sun drives the wind and ocean currents. Also, it is amazing that in less than 2 h, the energy from the sun striking the Earth's surface, if captured, could supply all the world's energy needs for a year (The Economist, 2016).

We've seen from the previous tables in this chapter how solar voltaic power has increased over the past 10 years. It's clear that the future for solar power is very promising. Solar power had $161 billion in new investment in 2015, more than natural gas and coal combined. The solar energy movement that began in northern Europe has now been growing rapidly in countries where power needs are growing fast and there is plenty of sun. For the first time in 2015, developing countries attracted more investment in renewable energy than rich ones. Poorer countries, from China to Chile, are increasingly getting their electricity from giant solar parks in arid places linked to their national grids. China, the new world leader in solar power, and India each plan to add about 100 GW in the next 4 and 6 years, respectively.

Lower costs help explain this extraordinary expansion. The price of solar panels, which are produced almost exclusively in China, has fallen by 80% in the past 5 years. A new business model is proving just as beneficial. Grid providers are offering long-term contracts to private firms to produce large amounts of solar energy, which in turn helps those firms secure cheap finance and cut prices. A recent tender in Mexico will generate electricity at a record low cost of $40 a megawatt hour – cheaper than natural gas or coal.

The sun provides power when it is needed most, during daylight hours when air-conditioning systems are running at full blast. And by reducing the need to import carbon-burning fossil fuels, solar power helps the planet as well as the balance of payments in such countries.

Harnessing the Sun

Of course, some caution is warranted. Bids to provide solar power may prove to be too ambitious. Two failed renewable-energy providers, America's SunEdison and Spain's Abengoa, provide lessons on the dangers of financial engineering and taking on too much debt in order to expand quickly.

Bottlenecks in energy infrastructure are another problem. Developing countries will need to invest more in building transmission lines to connect the solar power being generated in far-flung deserts with its users. Makers of solar panels should focus not just on slashing their cost but on improving the technology so that more of the sun's energy is converted into electrical power. The intermittency of the sun will remain an issue. But if storage costs continue to decline, the possibility of combining batteries on land with energy from the sun will be an excellent solution to our planet's long-term renewable energy needs.

Appendix C. Worldwide Average Prices per kWh for Electricity impacting Healthcare Costs

Green is not simply a new form of generating electric power.
It is a new form of generating national power – period.
　　　　　　　—Thomas L. Friedman – Author of "Hot, Flat, and Crowded"

Electricity costs have a significant impact on the location of new data centers. Also, as discussed throughout this book, electricity cost can also be a major cost factor for hospitals and healthcare facilities. This appendix gives information on how the cost of electricity varies around the world and within countries such as the United States.

Of course the cost of electricity goes beyond the data center in its impact on green IT. Thomas Friedman, in his book *Hot, Flat and Crowded: Why We Need a Green Revolution and How It Can Renew America*, (Friedman, 2008), sees the fusion of IT and energy technologies, envisioning all the power systems in your home as communicating with all the information systems in your home and that they had all merged into one big seamless platform for using, storing, generating, and even buying and selling clean energy. Friedman sees the need to have information technology and energy technology, IT and ET, merged into a single system that he calls an Energy Internet (ET). He envisions how professionals with IT skills will be needed to build integral components of this clean-energy web. Also, as seen in Appendix B, electricity generation from renewable sources, such as wind and solar, has significantly come down in cost per kWh over the past few years. Electricity generation from renewable sources is becoming the least expensive way to generate electricity, worldwide.

Getting back to healthcare data centers, here's an example of how the cost of electricity can have an impact on the location of data centers. If the cost of electricity is 22 cents/kWh in New York City and only 4.5 cents/kWh in Charleston, West Virginia, then the location of a large new data center in West Virginia could potentially save millions of dollars in electricity costs each year. With high-speed networks (including the Internet) available all over the United States and, for a large part, all over the world, data centers can now be located almost anywhere. So, these days, hospitals and healthcare facilities can access their data centers easily from almost anywhere. And the location of the data center can be influenced by electricity costs and other factors. The hospitals and healthcare centers, themselves, need to be located in areas that are convenient for their patients.

© Springer International Publishing AG, part of Springer Nature 2018
N. S. Godbole, J. P. Lamb, *Making Healthcare Green*,
https://doi.org/10.1007/978-3-319-79069-5

As indicated in this appendix, the cost per kWh and emission estimates can vary widely on a worldwide basis. This appendix is intended to give information on how the cost of electricity varies around the world – and how this could impact the location of new data centers.

US Average Electricity Prices by State

US Department of Energy EIA (Energy Information Agency) released January 2017.

http://www.eia.doe.gov/cneaf/electricity/st_profiles/e_profiles_sum.html

Electrical prices can vary significantly within the United States. Based on the following table, average electrical prices per kWh during 2015 range from as low as 7.65 cents/kWh in Louisiana to as high as 26.17 cents/kWh in Hawaii. The average US price per kWh during 2017 was 10.41 cents/kWh.

State Electricity Profiles

Data for 2015 | Release Date: January 17, 2017

Table C.1   US average electricity prices by state (2015)

Name	Average retail price (cents/kWh)	Net summer capacity (MW)	Net generation (MWh)	Total retail sales (MWh)
Alabama	9.33	30,407	152,477,427	88,845,543
Alaska	17.59	2589	6,284,937	6,159,204
Arizona	10.34	28,077	113,142,048	77,349,416
Arkansas	8.19	14,707	55,559,428	46,465,154
California	15.42	74,892	196,703,858	261,170,437
Colorado	9.94	15,793	52,393,077	54,116,046
Connecticut	17.77	8784	37,470,622	29,476,155
Delaware	11.17	3403	7,810,006	11,498,205
District of Columbia	12.07	21	53,750	11,291,233
Florida	10.49	58,636	237,412,633	235,599,398
Georgia	9.62	36,303	128,817,898	135,878,215
Hawaii	26.17	2674	10,119,500	9,511,352
Idaho	8.09	4948	15,667,095	23,058,814
Illinois	9.4	45,532	193,952,040	138,619,970
Indiana	8.99	26,324	104,019,275	104,514,518
Iowa	8.35	16,854	56,658,918	47,147,293

(continued)

Table C.1   (continued)

Name	Average retail price (cents/kWh)	Net summer capacity (MW)	Net generation (MWh)	Total retail sales (MWh)
Kansas	10.14	14,350	45,527,124	39,849,127
Kentucky	8.14	20,061	83,543,671	76,038,630
Louisiana	7.65	26,235	107,812,354	91,676,489
Maine	12.78	4615	11,741,265	11,888,168
Maryland	12.07	12,408	36,365,544	61,781,719
Massachusetts	16.9	13,236	32,085,969	54,621,088
Michigan	10.76	30,063	113,008,050	102,479,921
Minnesota	9.53	15,743	56,979,768	66,579,234
Mississippi	9.53	16,085	64,757,864	48,691,529
Missouri	9.44	21,764	83,640,067	81,504,081
Montana	8.9	6180	29,302,401	14,206,911
Nebraska	8.91	8658	39,883,391	29,495,073
Nevada	9.48	10,878	39,046,784	36,019,690
New Hampshire	16.02	4438	20,015,893	10,999,149
New Jersey	13.74	18,767	74,608,860	75,489,623
New Mexico	9.62	8404	32,701,398	23,093,553
New York	15.28	40,249	138,627,721	148,913,655
North Carolina	9.37	31,310	128,388,445	133,847,523
North Dakota	8.75	7362	37,156,612	18,128,948
Ohio	9.98	28,711	121,893,401	149,213,224
Oklahoma	7.9	24,835	76,135,596	61,336,385
Oregon	8.75	15,916	57,866,535	47,263,974
Pennsylvania	10.31	42,344	214,572,291	146,344,028
Rhode Island	17.01	1849	6,939,019	7,664,718
South Carolina	9.58	22,698	96,532,213	81,328,246
South Dakota	9.47	4126	9,633,033	12,101,979
Tennessee	9.3	20,220	75,214,636	99,632,108
Texas	8.7	117,144	449,826,336	392,337,354
Utah	8.54	8329	41,949,120	30,192,350
Vermont	14 41	655	1,982,047	5,521,109
Virginia	9.31	25,182	84,411,592	112,009,045
Washington	7.4	31,003	109,287,458	90,116,086
West Virginia	8.11	15,115	72,295,269	32,303,026
Wisconsin	10.73	16,669	66,360,183	68,698,932
Wyoming	7.97	8512	48,966,519	16,924,762
U.S. Total	10.41	1,064,055	4,077,600,939	3,758,992,390

Worldwide Electricity Prices for Industry by Country

Table C.2 below shows simple comparison of electricity tariffs in industrialized countries and territories around the world, expressed in US dollars. The comparison does not take into account factors including fluctuating international exchange rates, a country's purchasing power, government electricity subsidies, or retail discounts that are often available in deregulated electricity markets. The link is:

https://en.wikipedia.org/wiki/Electricity_pricing

For example, in 2015 (see Table C.1), Hawaii residents had the highest average residential electricity rate in the United States (26.17¢/kWh), while Louisiana residents had the lowest average residential electricity costs (7.65¢/kWh). Even in the contiguous United States, the gap is significant, with Connecticut residents having the highest average residential electricity rates in the lower 48 US states (17.77¢/kWh).

Forecasting

Electricity price forecasting is the process of using mathematical models to predict what electricity prices will be in the future.

Forecasting Methodology

The simplest model for day-ahead forecasting is to ask each generation source to bid on blocks of generation and choose the cheapest bids. If not enough bids are submitted, the price is increased. If too many bids are submitted, the price can reach zero or become negative. The offer price includes the generation cost as well as the transmission cost, along with any profit. Power can be sold or purchased from adjoining power pools.

Wind and solar power are non-dispatchable. Such power is normally sold before any other bids, at a predetermined rate for each supplier. Any excess is sold to another grid operator, stored, using pumped-storage hydroelectricity, or, in the worst case, curtailed. The HVDC Cross-Channel line between England and France is bidirectional but is normally used to purchase power from France. Allocation is done by bidding.

Driving Factors

In addition to production costs, electricity prices are set by supply and demand. Everything from salmon migration to forest fires can affect power prices. However, some fundamental drivers are the most likely to be considered.

Table C.2 Worldwide electricity prices for industry by country (recent dates given)

Country/territory	US cents/kWh	Date
Argentina (Concordia)	19.13[a]	Jun 14, 2013
Australia	Varies by state anywhere from 15 to 26 per kWh	Dec 21, 2016
Bahrain	0.79–4.23	Aug 19, 2015
Bangladesh	2.95–9.24	Mar 13, 2014
Belarus	13.8–69.8	Jun 21, 2016
Belgium	29.08	Nov 1, 2011
Brunei	0.72–8.64	Jun 18, 2017
Brazil	12.00–25.00 varying by state and electricity service provider	Jul 7, 2016
Cambodia	15.63–21.00 in Phnom Penh	Feb 28, 2014
Canada, Ontario, Toronto	6.52–11.69 depending on time of day plus transmission, delivery, and other charges of about 3.75 per kWh	Feb 9, 2014
Canada, Quebec	4.60 for the first 33 kWh/day then 7.05 + 32.14/day for subscription fee (all converted to USD on July 17, 2017)	2017
China	4–4.5	2014
Chile	23.11	Jan 1, 2011
Colombia (Bogota)	18.05	Jun 1, 2013
Cook Islands	34.6–50.2	
Croatia	17.55	Jul 1, 2008
Curaçao	26.58–35.08	Aug 1, 2017
Denmark	33	May 1, 2015
United Arab Emirates – Dubai	6.26–10.35 (plus 1.63 fuel surcharge)	
United Arab Emirates – Abu Dhabi	0–8.23 (i.e., AED 0 to AED 0.305)	2017
Egypt	Priced into sections at a kWh/month, subsidized[a]	Jul 17, 2014
	0.98 @ 0–50 kWh/M	
	1.89 @ 51–100 kWh/M	
	2.08 @ 0–200 kWh/M	
	3.12 @ 201–350 kWh/M	
	4.42 @ 351–650 kWh/M	
	7.8 @ 651–1000 kWh/M	
	9.62 @ 1000+ kWh/M	
Ethiopia	6.7–7.7[a]	Dec 31, 2012
Fiji	12–14.2	
Finland	20.65	Nov 1, 2011
France	19.39	Nov 1, 2011
Georgia	8.00	Jul 24, 2015
Germany	35.00	Mar 1, 2017
Romania	18.40	Jun 26, 2013
Guyana	26.80	Apr 1, 2012

(continued)

Table C.2 (continued)

Country/territory	US cents/kWh	Date
Switzerland	25.00	Jan 6, 2014
Hungary	23.44	Nov 1, 2011
Hong Kong	12.04–24.05	Jan 1, 2013
India	0.1–18 (average 7)	March 1, 2014
Indonesia	11	Jul 21, 2015
Iceland	5.54	Nov 8, 2015
Iran	2–19	Jul 1, 2011
Ireland	23.06	November 1, 2016
Israel	15.35[a]	May 8, 2017
Italy	28.39	Nov 1, 2011
Jamaica	44.7	Dec 4, 2013
Japan	20–24	Dec 31, 2009
Jordan	5–33[a]	Jan 30, 2012
Kazakhstan	4.8–8.2	Dec 13, 2016
Kiribati	32.7	
Kuwait	0.3–3	Jan 1, 2016
Laos	11.95 for >150 kWh, 4.86 for 26–150 kWh, 4.08 for 0–25 kWh	Feb 28, 2014
Latvia	18.23 calculated for 100 kWh including transport, green energy tax and VAT	Jun 14, 2017
Lithuania	12	July 1, 2016
Macedonia	7–10	Aug 1, 2013
	industrial-14	
Malaysia	Domestic Consumer pricing per kWh used, subsidized	Jan 1, 2014
	4.95 @ 1–200 kWh	
	7.59 @ 201–300 kWh	
	11.73 @ 301–600 kWh	
	12.41 @ 601–900 kWh	
	12.98 @ 901 kWh onwards	
	(Exchange rate of 4.4 MYR to 1 USD on Nov 24, 2016)	
Marshall Islands	32.6–41.6	
Mexico	19.28[b]	Aug 22, 2012
Moldova	11.11	Apr 1, 2011
Myanmar	3.6	Feb 28, 2014
Nepal	7.2–11.2	Jul 16, 2012
Netherlands	28.89	Nov 1, 2011
New Caledonia	26.2–62.7	
New Zealand	19.15	Apr 19, 2012
Niue	44.3	
Nigeria	2.58–16.55	Jul 2, 2013

(continued)

Table C.2 (continued)

Country/territory	US cents/kWh	Date
Norway	15.9	Jul 25, 2013
Pakistan	General supply tariff – residential	Jul 14, 2015
	2 <50 kWh/M	
	5.79 @ 1–100 kWh/M	
	8.11 @ 101–200 kWh/M	
	10.21 @ 201–300 kWh/M	
	16 @ 301–700 kWh/M	
	18 >700 kWh/M	
Palau	22.83	
Papua New Guinea	19.6–38.8	
Peru	10.44	2007
Philippines	18.22	October 7, 2015
	Palawan 25.2	
Portugal	25.25	Nov 1, 2011
Russia	2.4–14	Nov 1, 2011
Rwanda	22–23.6	2016
Saudi Arabia	1–7 (from the first 2000 kWh/month to more than 10,000 kWh/month)	Sep 9, 2015
Serbia	3.93–13.48, average ~6,1[c]	Feb 28, 2013
Singapore	20.30	Jun 16, 2017
Spain	19.72	June 2017
	(21% VAT +6% electricity tax are included in this price)	
Solomon Islands	88–99	
South Africa	15	Sep 29, 2015
Surinam	3.90–4.84	Nov 20, 2013
Sweden	8.33	Feb 3, 2015
Tahiti	25–33.1	
Thailand	6–13	July 1, 2013
Tonga	47	Jun 1, 2011
Trinidad and Tobago	4	July 8, 2015
Turkey	11.20 residential (low voltage)	Jul 1, 2016
	11.29 business (low voltage)	
	8.78 industry (medium voltage)	
Ukraine	2.6–10.8	2017
United Kingdom	22	May 1, 2015
United States	8–17; 37[d] 43[d]	Sep 1, 2012
United States Virgin Islands	48.9–51.9	Oct 1, 2014
Uruguay	17.07–26.48	Feb 11, 2014
Uzbekistan	4.95	2011
Vanuatu	60	

(continued)

Table C.2   (continued)

Country/territory	US cents/kWh	Date
Venezuela	0.016 at commonly used unofficial exchange rate (3684 VEF/USD) or 0.089 cents at official exchange rate (678 VEF/USD)	2016-12-01
Vietnam	6.20–10.01	2011
Western Samoa	30.5–34.7	

The US Energy Information Administration (EIA) also publishes an incomplete list of international energy prices, while the International Energy Agency (IEA) provides a thorough, quarterly review

[a]Denotes countries with government subsidized electricity tariffs
[b]Mexico subsidizes electricity according to consumption limits. More than 500 kWh consumed bimonthly receive no subsidies. Only 1% of Mexico's population pays this tariff
[c]Prices don't include VAT (20%)
[d]Hawaii

Power Quality

Transmission, production, and consuming electrical power associated with excessive total harmonic distortions (THD) and not unity power factor (PF) would be costly for owners. Cost of PF and THD impact is difficult to estimate, but it causes heat and vibration, malfunctioning, and even meltdowns. The electric company monitors the transmission level. A spectrum of compensation devices mitigates bad outcomes, but improvements can be achieved only with real-time correction devices (old-style switching type, modern low-speed DSP driven, and near real-time). Most modern devices reduce problems while maintaining return on investment and significant reduction of ground currents. Another reason to mitigate the problems is to reduce operation and generation costs, which is commonly done by electric power distribution companies in conjunction with generation companies. Power quality problems can cause erroneous responses from many kinds of analog and digital equipment, where the response could be unpredictable.

Phase Balancing

Most common distribution network and generation is done with three phase structures, with special attention paid to the phase balancing and resulting reduction of ground current. It is true for industrial or commercial networks where most power is used in three phase machines, but light commercial and residential users do not have real-time phase balancing capabilities. Often this issue leads to unexpected equipment behavior or malfunctions and, in extreme cases, fires. For example, sensitive professional analogue or digital recording equipment must be connected to well-balanced and grounded power networks. To determine and mitigate the cost of the unbalanced electricity network, electric companies in most cases charge by demand or as a separate category for heavy unbalanced. A few simple techniques are available for balancing that require fast computing and real-time modeling.

Weather

Studies show that generally demand for electricity is driven largely by temperature. Heating demand in the winter and cooling demand (air conditioners) in the summer are what primarily drive the seasonal peaks in most regions. Heating degree days and cooling degree days help measure energy consumption by referencing the outdoor temperature above and below 65 °F, a commonly accepted baseline.

Hydropower Availability

Snowpack, stream flows, seasonality, salmon, etc. all affect the amount of water that can flow through a dam at any given time. Forecasting these variables predicts the available potential energy for a dam for a given period. Some regions such as Egypt, China, and the Pacific Northwest get significant generation from hydroelectric dams.

Power Plant and Transmission Outages

Whether planned or unplanned, outages affect the total amount of power that is available to the grid.

Fuel Prices

The fuel used to generate electricity is the primary cost incurred by electrical generation companies. This will change as more renewable energy is used. Capital costs are the primary cost of solar and wind energy because they have no fuel cost.

Economic Health

During times of economic hardship, many factories cut back production due to a reduction of consumer demand and therefore reduce production-related electrical demand.

Projections for Worldwide Clean Energy Cost Comparisons

Table C.2 gives an indication of worldwide electricity prices for industry by country for recent dates. So what are projections for the future? Each of us knows from personal experience that electricity costs keep going up. However, costs for renewable/sustainable power generation are projected to continue to go down. Appendix B includes information on the continuing significant progress in reducing costs for electric generation by wind power and solar power. Hydro is currently the source of renewable energy making the most substantial contributions to global energy demand (Table B.1). However, hydro plants clearly have limited growth potential and can significantly damage aquatic ecosystems. Other traditional power generation methods such as nuclear are seeing a resurgence. Nuclear, although clean from a CO2 perspective, will continue to bring up concerns on safe disposable of spent fuel and overall safety concerns. From a cost per kWh standpoint, however, nuclear power generation has definite advantages. This section looks at projections for costs to generate electricity from nuclear, wind, and solar power.

Bloomberg New Energy Finance

https://about.bnef.com/blog/global-wind-solar-costs-fall-even-faster-coal-fades-even-china-india/

Global wind and solar costs to fall even faster, while coal fades even in China and India

June 15, 2017

This year's forecast from BNEF sees solar energy costs dropping a further 66% by 2040 and onshore wind by 47%, with renewables undercutting the majority of existing fossil power stations by 2030.

London and New York, June 15, 2017 – Renewable energy sources such as solar and wind are set to take almost three quarters of the $10.2 trillion the world will invest in new power generating technology over the years to 2040, according to a major independent forecast published today.

New Energy Outlook 2017, the latest long-term forecast from Bloomberg New Energy Finance, shows earlier progress than its equivalent a year ago toward decarbonization of the world's power system – with global emissions projected to peak in 2026 and to be 4% lower in 2040 than they were in 2016.

This year's report suggests that the greening of the world's electricity system is unstoppable, thanks to rapidly falling costs for solar and wind power, and a growing role for batteries, including those in electric vehicles, in balancing supply and demand. Seb Henbest, lead author of NEO 2017 at BNEF

NEO 2017 is the result of 8 months of analysis and modeling by a 65-strong team at Bloomberg New Energy Finance. It is based purely on the announced project pipelines in each country, plus forecast economics of electricity generation and power system dynamics. It assumes that current subsidies expire and that energy policies around the world remain on their current bearing.

Here are some key findings from this year's forecast:

- Solar and wind dominate the future of electricity. We expect $7.4 trillion to be invested in new renewable energy plants by 2040 – which is 72% of the $10.2 trillion that is projected to be spent on new power generation worldwide. Solar takes $2.8 trillion and sees a 14-fold jump in capacity. Wind draws $3.3 trillion and sees a fourfold increase in capacity. As a result, wind and solar will make up 48% of the world's installed capacity and 34% of electricity generation by 2040, compared with just 12% and 5% now.

- Solar energy's challenge to coal gets broader. The levelized cost of electricity from solar PV, which is now almost a quarter of what it was just in 2009, is set to drop another 66% by 2040. By then a dollar will buy 2.3 times as much solar energy than it does today. Solar is already at least as cheap as coal in Germany, Australia, the United States, Spain, and Italy. By 2021, it will be cheaper than coal in China, India, Mexico, the United Kingdom, and Brazil as well.
- Onshore wind costs fall fast, and offshore falls faster. Offshore wind levelized costs will slide a whopping 71% by 2040, helped by development experience, competition and reduced risk, and economies of scale resulting from larger projects and bigger turbines. The cost of onshore wind will fall 47% in the same period, on top of the 30% drop of the past 8 years, thanks to cheaper, more efficient turbines and streamlined operating and maintenance procedures.

Glossary

AEM (Active Energy Manager) An IBM tool to measure energy use on each server and other IT devices.

AFCOM (Association for Computer Operations Management) An organization that provides education and resources for data-center managers.

AIX (Advanced Interactive Executive) AIX is IBM's version of UNIX

Alliance to Save Energy An organization that promotes energy efficiency worldwide.

AMD (Advanced Micro Devices) A microprocessor manufacturer and the main competitor to Intel

Analytics The discovery, interpretation, and communication of meaningful patterns in data.

APC (American Power Conversion) A company that deals with data-center efficiency.

ASHRAE American Society of Heating, Refrigerating, and Air Conditioning Engineers

Bandwidth The amount of data that can be transmitted across a particular network. Basic Ethernet has a 10 Mbps bandwidth; however 100 Mbps Ethernet and recently Gigabit Ethernet are common for corporate LANs and server farm infrastructure.

Big Data A term for data sets that are so large or complex that traditional data processing applications are inadequate to deal with them.

BIOS In computing, BIOS is an acronym that stands either for the basic input/output system or for built-in operating system. BIOS usually refers to the firmware code run by a PC when first powered on. The primary function of the BIOS is to identify and initialize system component hardware (such as the video display card, hard disk, and floppy disk) and some other hardware devices.

Blade server A chassis housing that contains multiple, modular electronic circuit boards (blades), each of which includes processors, memory, storage, and network connections and can act as a server on its own. The thin blades can be added or removed, depending on needs for capacity, power, cooling, or networking traffic.

© Springer International Publishing AG, part of Springer Nature 2018
N. S. Godbole, J. P. Lamb, *Making Healthcare Green*,
https://doi.org/10.1007/978-3-319-79069-5

BPR Business Process Reengineering. BPR is the analysis and redesign of work-flows within and between enterprises in order to optimize end-to-end processes and automate non-value-added tasks.

bps (bits per second) The rate of data transmission across a network.

Browser See Web Browser

BTU British Thermal Unit. The BTU is a unit of energy used in the power, steam generation, and heating and air-conditioning industries. A BTU is defined as the amount of heat required to raise the temperature of 1 lb of liquid water by 1 °F.

CAGR Compound annual growth rate.

Carbon footprint The total amount of greenhouse gases produced to directly and indirectly support human activities, usually expressed in equivalent tons of carbon dioxide (CO_2).

CDU Cabinet power distribution unit (CDU) is an intelligent power distribution unit (PDU) with local input current monitoring to allow precise measurement of the electric current (in amps) that network devices (e.g., computer room air-conditioning units, servers, etc.) are drawing on the power circuit.

CEO Chief Executive Officer

CERN The European Organization for Nuclear Research (acronym, originally from the French: Conseil Européen pour la Recherche Nucléaire). CERN is the world's largest particle physics laboratory, situated in the northwest suburbs of Geneva on the Franco-Swiss border. CERN has need for much high performance computing and represents a real and a virtual workplace for many scientists and engineers, representing 500 universities and 80 nationalities.

CHP Combined heat and power. Combined heat and power (CHP) plants recover otherwise wasted thermal energy for heating. The supply of high-temperature heat first drives a gas or steam turbine-powered electric generator. The resulting low-temperature waste heat is then used for water or space heating.

CIO Chief information officer

CLEER model Cloud Energy and Emission Research model. The CLEER model is an open access model for accessing the net energy and emission implications of cloud services at different levels.

Climate crisis The adverse long-term effects of global climatic changes have been named the climate crisis and pose a very real and ongoing threat, not only to our continued existence as a species but also to the Earth's ability to support life.

Cloud computing The name used for a subset of grid computing that includes utility computing and other approaches to the use of shared computing resources. Cloud computing is an alternative to having local servers or personal devices handling users' applications. Essentially, it is an idea that the technological capabilities should "hover" over everything and be available whenever a user wants.

Cogeneration The use of a heat engine or a power station to simultaneously generate both electricity and useful heat. In the United States, Con Edison distributes 30 billion pounds of 350 °F/180 °C steam each year through its seven cogeneration plants to 100,000 buildings in Manhattan – the biggest steam district in the world.

CPU Central processing unit. A CPU is the "brain" of a computer.

CRAC Computer room air conditioners. These devices control humidity through humidification and dehumidification as required, both of which consume energy.

Cricket A system for monitoring trends in time-series data, initially developed to help network managers visualize and understand the traffic on their networks (http://cricket.sourceforge.net/).

CRM Customer relationship management. CRM is a term that refers to practices, strategies, and technologies that companies use to manage and analyze customer interactions and data throughout the customer lifecycle.

DAS Direct attached storage.

Data at rest Data at rest generally refers to data stored in persistent storage (disk, tape).

Data centers Facilities that primarily contain electronic equipment used for data processing, data storage, and communication networking.

Data deduplication Often called intelligent compression or single-instance storage, it is a process that eliminates redundant copies of data and reduces storage overhead. Data deduplication techniques ensure that only one unique instance of data is retained on storage media, such as disk, flash, or tape.

Data science An interdisciplinary field about processes and systems to extract knowledge or insights from data in various forms, either structured or unstructured.

DCiE Data center infrastructure efficiency. The DCiE = (IT equipment power × 100%)/total facility power. This is the reciprocal of the Green Grid's PUE (power usage effectiveness). The DCiE is considered to be a more intuitive metric than the PUE since the DCiE can range from 0% to 100% where 100% would be the ideal efficiency. The PUE can range from 1 to a very large number, where 1 is considered ideal. So a PUE of 1.1 is very good, while a PUE above 3 (a DCiE below 33%) is considered very inefficient.

DG Distributed generation. This would include fuel cells and other clean, efficient distributed technologies used in data centers.

DOE Department of Energy. A US governmental department with a mission to advance the national, economic, and energy security of the United States and to promote scientific and technological innovation in support of that mission.

DOE DC Pro Tool Department of Energy Data Center Profiler Tool. An online software tool designed to help industries worldwide quickly "diagnose" how energy is being used by their data centers and how they might save energy and money. The tool is available at http://www1.eere.energy.gov/industry/saveenergynow/printable_versions/partnering_data_centers.html.

EBM Evidence-based medicine. EBM is the conscientious, explicit, judicious, and reasonable use of modern, best evidence in making decisions about the care of individual patients.

EDUCAUSE A nonprofit association whose mission is to advance higher education by promoting the intelligent use of information technology (http://www.educause.edu/).

EEC European Energy Community. A regulatory framework for trading energy in Europe.

eHealth A relatively recent term for healthcare practice supported by electronic processes and communication, dating back to at least 1999.

EHR Electronic health record. An electronic health record (EHR) is a digital version of a patient's paper chart. EHRs are real-time, patient-centered records that make information available instantly and securely to authorized users.

EMR Electronic medical record. An electronic medical record (EMR) is a digital version of the traditional paper-based medical record for an individual. The EMR represents a medical record within a single facility, such as a doctor's office or a clinic. The difference between an EHR and an EMR varies greatly, although many use the terms EHR and EMR interchangeably. As providers are slowly becoming mandated across the country to use electronic records, there is an intense amount of news coverage covering electronic health records (EHR) and electronic medical records (EMR). The most basic difference is that an electronic medical record (EMR) is a digital version of a chart with patient information stored in a computer, not a filing cabinet, and an electronic health record (EHR) is a digital record of health information.

Ensemble A pool of homogeneous systems within a grid or cloud computer system which are compatible with one another.

EPA Environmental Protection Agency. A US governmental agency with a mission to protect human health and the environment.

EPEAT Electronic product environmental assessment tool. EPEAT was created through an Institute of Electrical and Electronics Engineers (IEEE) council because companies and government agencies wanted to put green criteria in IT requests for proposals. EPEAT's energy-consumption criteria are based on the EPA's Energy Star requirements for PCs, and the "sensitive material" criteria require companies to meet the European Union's tough standards for limiting the hazardous chemicals and components used to make them.

Epidemiology The study and analysis of the patterns, causes, and effects of health and disease conditions in defined populations. It is the cornerstone of public health and shapes policy decisions and evidence-based practice by identifying risk factors for disease and targets for preventive healthcare.

ESG Enterprise Strategy Group. ESG is an IT analyst and consulting firm focused on information storage, security, and management.

ESPC Energy services performance contract. An incentive envisioned by the US federal government for efficiency upgrades for data centers.

EUI Energy use intensity. The EPA uses EUI (kBTU/square foot) to determine building Energy Star ratings. See www.energystar.gov for the range of EUI Energy Star (green buildings) ranges depending on building type.

eWaste Electronic waste.

Exabyte The exabyte is a multiple of the unit byte for digital information. In the International System of Units (SI), the prefix exa indicates multiplication by the sixth power of 1000 (1018). Therefore, one exabyte is one quintillion bytes (short scale). The symbol for the exabyte is EB.

Fiber channel The technology commonly used to connect server to external data storage (SAN).

FIM Federated identity management. An arrangement that can be made among multiple enterprises that lets subscribers use the same identification data to obtain access to the networks of all enterprises in the group.

GB (gigabyte) A billion bytes of computer or hard disk memory.

GDP Gross domestic product.

Genomic data Genomic data refers to the genome and DNA data of an organism. They are used in bioinformatics for collecting, storing, and processing the genomes of living things. Genomic data generally require a large amount of storage and purpose-built software to analyze.

GGHH Global Green and Healthy Hospitals. GGG has 886 members in 49 countries on 6 continents who represent the interests of over 28,000 hospitals and health centers.

GHG Greenhouse gas. A greenhouse gas (abbrev. GHG) is a gas in an atmosphere that absorbs and emits radiation within the thermal infrared range. This process is the fundamental cause of the greenhouse effect. The primary greenhouse gases in Earth's atmosphere are water vapor, carbon dioxide, methane, nitrous oxide, and ozone. Without greenhouse gases, the average temperature of Earth's surface would be about $-18\,°C$ ($0\,°F$), rather than the present average of $15\,°C$ ($59\,°F$).

GIPC (Green IT Promotion Council) An organization established in Japan in 2008 to address global warming by electronics firms, related industry bodies, and other groups.

Green Grid See "The Green Grid."

Greenhouse gas A greenhouse gas (abbrev. GHG) is a gas in an atmosphere that absorbs and emits radiation within the thermal infrared range. See GHG.

Green IT Green information technology

Greenpeace A group originally founded in Vancouver, British Columbia, Canada, in 1971 to oppose the US testing of nuclear devices in Alaska. The focus of the organization later turned to other environmental issues: whaling, bottom trawling, global warming, old growth, and nuclear power. Greenpeace has national and regional offices in many countries. They also have a big presence worldwide, all of which are affiliated to the Amsterdam office of Greenpeace International.

Green washing Projects or processes that appear to be more green than they are. Similar concept to "white washing."

Grid computing A major evolutionary step that virtualizes an IT infrastructure. It's defined by the Global Grid Forum (www.gridforum.org) as distributed computing over a network of heterogeneous resources across domain boundaries and enabled by open standards.

GUI (Graphical user interface) A pictorial way of representing to a user the capabilities of a system and the work being done on it.

Hadron Collider The Large Hadron Collider (LHC) is the world's largest and highest-energy particle accelerator. Built by CERN, the LHC drives much of the need for performance computing (HPC) at hundreds of universities and laboratories around the world.

Healthcare The maintenance or improvement of health via the diagnosis, treatment, and prevention of disease, illness, injury, and other physical and mental impairments in human beings.

HIPAA Health Insurance Portability and Accountability Act. HIPPA of 1996 (HIPAA; Pub.L. 104–191, 110 Stat. 1936, enacted August 21, 1996) was enacted by the US Congress and signed by President Bill Clinton in 1996.

HIT Health information technology

Host In the TCP/IP sense, a computer that allows users to communicate with other host computers on a network. Individual users communicate by using application programs, such as electronic mail and FTP. Also used to refer to a large computer system, such as a mainframe.

HPC High-performance computing. HPC uses supercomputers and computer clusters to solve advanced computation problems.

HTML (hypertext markup language) This is the language used to write World Wide Web documents or pages. It is a subset of ISO SGML.

HTTP (hypertext transfer protocol) This is the protocol used by the World Wide Web to transfer documents between clients and servers.

HVAC Heating, ventilating, and air conditioning.

Hyper-V This is a Microsoft Windows Server 2008 hypervisor-based server virtualization technology. Allows separate virtual machines (VMs) running on a single physical machine. Like other virtualization software (e.g., VMware), Hyper-V also can run multiple different operating systems – Windows, Linux, and others – in parallel, on a single server.

Hypervisor A hypervisor, also called a virtual machine manager, is a program that allows multiple operating systems to share a single hardware host. Each operating system appears to have the host's processor, memory, and other resources all to itself.

IaaS Infrastructure as a service. IaaS is a form of cloud computing that provides virtualized computing resources over the Internet.

ICT Information and communication technology.

IDC International Data Corporation. A market research and analysis firm specializing in information technology, telecommunications, and consumer technology.

Information and Communication Technology (ICT) An extended term for information technology (IT) which stresses the role of unified communications and the integration of telecommunications (telephone lines and wireless signals), computers, as well as necessary enterprise software, middleware, storage, and audio-visual.

Information Technology (IT) The application of computers to store, study, retrieve, transmit, and manipulate data, or information, often in the context of a business or other enterprises. IT is considered a subset of information and communication technology (ICT).

INR Indian rupees.

Insight control A tool from HP for measuring and managing energy use in servers and other IT equipment. Insight control allows management of HP ProLiant and BladeSystem infrastructure systems. The management includes power measurement and power capping.

Intel The major microprocessor manufacturer of CPUs used in Windows PCs and servers.

Internet A set of connected networks. The term Internet refers to the large and growing public domain internet developed by DARPA that uses TCP/IP. It is shared by universities, corporations, and private individuals.

Internet of Things (IoT) The network of physical devices, vehicles, home appliances, and other items embedded with electronics, software, sensors, actuators, and network connectivity which enable these objects to connect and exchange data.

Intranet A web network that connects computers within the same company or organization over a private network. An intranet offers higher security than an Internet web site because of the private nature of the network.

IP (Internet Protocol) The network-layer protocol for the Internet Protocol suite.

IT Information technology

ITIL Information Technology Infrastructure Library. ITIL is a set of concepts and policies for managing information technology (IT) infrastructure, development, and operations.

Ivy Plus technology consortium A consortium of top universities (Ivy League plus other universities such as MIT, Duke, and Stanford) that fosters collaboration to help drive future growth and use of technology.

KDD Knowledge discovery in databases. KDD is the process of discovering useful knowledge from a collection of data. This widely used data mining technique is a process that includes data preparation and selection, data cleansing, incorporating prior knowledge on data sets, and interpreting accurate solutions from the observed results.

KWH or kWh Kilowatt hour. A basic unit of electric energy equivalent to using power of 1000 W for 1 h. The kilowatt hour is the energy delivered by electric utilities that is usually expressed and charged for in kWh. Note that the kWh is the product of power in kilowatts multiplied by time in hours; it is not kW/h. Large industrial units are used for taking high voltage and current and reducing it to more common and useful levels, for example, from 240 V 30 A single phase to multiple 120 V 15 A or 120 V 20 A plugs. They are used in computer data centers, in stage shows, and in other electrically intensive applications

LED Light-emitting diode. LED lighting products produce light approximately 90% more efficiently than incandescent light bulbs.

LEED (Leadership in Energy and Environmental Design) is a green building rating system, developed by the US Green Building Council (USGBC), which provides a suite of standards for environmentally sustainable construction.

Liebert An American manufacturer of environmental, power, and monitoring systems for mainframe computer, server racks, and critical process systems. Liebert is part of the Emerson Corporation.

LPAR Logical partition. A logical partition, commonly called an LPAR, is a subset of a computer's hardware resources, virtualized as a separate computer. In effect, a physical machine can be partitioned into multiple logical partitions, each hosting a separate operating system.

mHealth (Also written as m-health or mobile health) is a term used for the practice of medicine and public health, supported by mobile devices.

mMedicine (mobile medicine) mMedicine provides useful medical tools and new communication platform between doctors and pharmaceutical companies.

MIB Management information base. A MIB stems from the OSI/ISO network management model and is a type of database used to manage the devices (such as routers and switches) in a communication network.

Mobile computing A technology that allows transmission of data, voice, and video via a computer or any other wireless-enabled device without having to be connected to a fixed physical link.

Moore's law Moore's law describes a long-term trend in the history of computing hardware. Since the invention of the integrated circuit in 1958, the number of transistors that can be placed inexpensively on an integrated circuit has increased exponentially, doubling approximately every 2 years. The trend was first observed by Intel cofounder Gordon E. Moore in a 1965 paper. It has continued for almost half of a century and is not expected to stop for another decade at least and perhaps much longer.

NAGIOS A popular open-source computer system and network monitoring application software. It watches hosts and services, alerting users when things go wrong and again when they get better. http://www.nagios.org/

NAS Network-attached storage.

NEDC New enterprise data center. The NEDC concept announced by IBM in 2008 is a vision for energy-efficient data centers based on lessons learned from working with customers on hundreds of data centers on best practice ways to approach energy efficiency for both existing and new data centers.

NYSERDA New York State Energy Research and Development Authority. NYSERDA is a public benefit corporation created in 1975. NYSERDA's earliest efforts focused solely on research and development with the goal of reducing the state's petroleum consumption. Subsequent research and development projects focused on topics including environmental effects of energy consumption, development of renewable resources, and advancement of innovative technologies (http://www.nyserda.org/).

NYSERNet New York State Education and Research network. This organization includes a shared data center in Syracuse, NY, members of the NYSGrid, member K–12 schools, colleges, universities, libraries, and corporate research labs.

NYSGrid New York State Grid. A New York State high performance computing (HPC) consortium.

OECD Organization for Economic Co-operation and Development.

OSI Open systems interconnection. The OSI reference model is an abstract description for layered communications and computer network protocol design.

PaaS Platform as a service. PaaS is a category of cloud computing services that provides a platform allowing customers to develop, run, and manage applications without the complexity of building and maintaining the infrastructure typically associated with developing and launching an app.

PDU Power distribution unit. A PDU is a device that distributes electric power.

Petawatt (PW) A petawatt is a unit of power equal to 10^{15} watts.

PPP Public-private partnership. A public-private partnership (PPP, 3P or P3) is a cooperative arrangement between two or more public and private sectors, typically of a long-term nature. Governments have used such a mix of public and private endeavors throughout history.

PSSC Products and Solutions Support Center. The IBM PSSC at Montpellier, France, is focused on benchmarking, performance, and sizing.

PUC Public utility commission

PUE Power usage effectiveness. PUE = total facility power/IT equipment power. A PUE of 1.5 for a data center is excellent (a green data center), while a PUE above 3.0 is considered quite inefficient. In order to make a more intuitive metric, the Green Grid defined the DCiE (data center infrastructure efficiency) that is the reciprocal of the PUE × 100%. So a DCiE of 100% is considered ideal, while a DCiE of 33% (a PUE of 3) is considered quite inefficient.

ROI Return on investment

RTO Recovery time objective

SaaS Software as a service. SaaS is a software licensing and delivery model in which software is licensed on a subscription basis and is centrally hosted.

SAN Storage area network. A storage area network provides access to consolidated, block-level data storage. SANs are primarily used to enhance storage devices, such as disk arrays, tape libraries, and optical jukeboxes, accessible to servers so that the devices appear to the operating system as locally attached devices.

Server A computer system that has been designated for running a specific server application or applications.

Server cluster A group of linked servers, working together closely so that in many respects they form a single server (computer). Clusters are usually deployed to improve performance and/or availability over that provided by a single server while typically being much more cost-effective than single servers of comparable speed or availability.

SMB Small- and medium-sized business.

SNMP Simple Network Management Protocol. SNMP is an Internet standard network monitoring protocol used in network management systems to monitor network-attached devices.

SPEC Standard Performance Evaluation Corporation.

STEaM STEM plus the arts.

STEM Science, technology, engineering, and mathematics.

Storage area network See SAN for definition.

Sustainability The US EPA defines sustainability as "meeting the needs of the present without compromising the ability of future generations to meet their own needs."

TCP/IP (Transmission Control Protocol/Internet Protocol) The set of applications and transport protocols that uses IP (Internet Protocol) to transmit data over a network. TCP/IP was developed by the Department of Defense to provide telecommunications for internetworking.

Telemedicine (Also referred to as "telehealth" or "e-health") Allows healthcare professionals to evaluate, diagnose, and treat patients in remote locations using telecommunication technology. Telemedicine allows patients in remote locations to access medical expertise quickly, efficiently, and without travel.

TERI The Energy and Resources Institute (TERI) is a research institute based in New Delhi that conducts research work in the fields of energy, environment, and sustainable development.

The Green Grid A consortium of more than 150 organizations working to improve data-center energy efficiency.

Ton (Air conditioning) One ton of air-conditioning capacity equals 12,000 BTUs per hour.

UAT User acceptance test

UESC Utility energy service contract. An incentive envisioned by the US federal government for efficiency upgrades for data centers.

UNIX The operating system originally designed by AT&T and enhanced by the University of California at Berkeley and others. Since it was powerful and essentially available for free, it became very popular at universities. Many vendors made their own versions of UNIX available, for example, IBM's AIX, based on OSF/1. The UNIX trademark and definition have since come under the control of X/Open, which will issue a unifying specification.

UPS Uninterruptible power supply. UPS, also known as a battery backup, provides emergency power. A UPS is typically used to protect computers, telecommunication equipment, or other electrical equipment where an unexpected power disruption could cause injuries, fatalities, serious business disruption, or data loss. A UPS can be used to provide uninterrupted power to equipment, typically for 5–15 min until a generator can be turned on or utility power is restored.

URL (universal resource locator) This is the World Wide Web name for a document, file, or other resources. It describes the protocol required to access the resource, the host where it can be found, and a path to the resource on that host.

ULTs Ultralow temperature freezers for lab and medical storage.

VFD Variable frequency drive. A VFD is a type of motor controller that drives an electric motor by varying the frequency and voltage supplied to the electric motor. This provides for energy efficiency.

Virtualization In computing, virtualization refers to the act of creating a virtual (rather than actual) version of something, including virtual computer hardware platforms, storage devices, and computer network resources.

VM Virtual machine. Virtual machine technology (such as Xen, VMWare, Microsoft Virtual Servers, the new Microsoft Hyper-V technology, etc.) enables multiple operating system environments to coexist on the same physical computer.

VMware A virtual server software product for Intel servers (Windows or Linux operating systems).

WAN (wide area network) A long-distance network for the efficient transfer of voice, data, and/or video between local, metropolitan, campus, and site networks. WANs typically use lower transfer rates (64 Kbps) or higher speed services such as T3, which operates at 45 Mbps. WANs also typically use common-carrier services (communication services available to the general public) or private networking through satellite and microwave facilities.

Watt A basic unit of electric power. Electric energy used is measured in kilowatt hours (kWh) which equates to using 1000 w of power for 1 h.

Web browser An application that provides an interface to the World Wide Web

Webmaster A person who manages a web site – similar to the network administrator

X86 A generic term that refers to the processor architecture commonly used in personal computers and servers. It derived from the model numbers, ending in "86," of the first few processor generations backward compatible with the original Intel 8086. The architecture has been implemented in processors from Intel, Cyrix, AMD, VIA, and many others.

Xen A free (open-source) software virtual machine monitor for IA-32, x86-64, IA-64, and PowerPC 970 architectures. It allows several guest operating systems to be executed on the same computer hardware at the same time.

Zettajoule (ZJ) A zettajoule is equal to one sextillion (10^{21}) joules.

References

Agrawal, B. (2013). Green cloud computing. *International Journal of Electronics and Communication Engineering & Technology (IJECET), 3*, 239–243.

Arthur, J. (2011). *Lean six sigma for hospitals*. New York: McGraw Hill.

Balaraman, P., & Kosalram, K. (2013). E –hospital management and hospital information systems – changing trends. *I.J. Information Engineering and Electronic Business, 5*, 50–58.

Becker. (2013, March 12). *16 top priorities for healthcare compliance programs in 2013*. Retrieved December 26, 2013, from Becker's Hospital Review: http://www.beckershospitalreview.com/legal-regulatory-issues/16-top-priorities-for-healthcare-compliance-programs-in-2013.html

Becker. (n.d.). *Four reasons why sustainability investments are vital to hospital finance strategies*. Retrieved from http://www.beckershospitalreview.com/racs-/-icd-9-/-10/4-reasons-why-sustainability-investments-are-vital-to-hospital-finance-strategies.html

Bishop, N. (2014, January 14). *Data quality and clinical coding for improvement*. Retrieved December 27, 2013, from HC-UK Conferences: http://www.healthcareconferencesuk.co.uk/nhs-data-quality-clinical-coding-training

Blake, H. (2008). Mobile phone technology in chronic disease management. *Nursing Standard, 23*(12), 43–46.

Bloomberg, M., & Pope, C. (2017). *Climate of hope: How cities, businesses, and citizens can save the planet*. New York: St. Martin's Press.

Boone, T. (2012). *Creating a culture of sustainability: Leadership, coordination and performance measurement decisions in healthcare*. Chicago: PRACTICE Greenhealth, UIC School of Public Health, Healthier Hospitals Initiatives, Healthcare without Harm.

Brannen, J. (n.d.). *Mixed methods research*. Retrieved from Institute of Education, University of London: http://eprints.ncrm.ac.uk/89/1/MethodsRevuewOaoerBCRM-005.pdf

Brant Simone, G. M. (2006). Access to care in rural China: A policy discussion. *International Economic Development Program*. The Gerald R. Ford School of Public Policy.

Brian, K. D. (2003). *Health reform in China: An analysis of rural health care delivery*. Armidale, NSW: University of New England, School of Economics.

Brookings Institution. (2014). *The use of mobile technology to improve health care and reduce costs in China and the United States*. Retrieved October 27, 2016, from BROOKINGS: https://www.brookings.edu/events/the-use-of-mobile-technology-to-improve-health-care-and-reduce-costs-in-china-and-the-united-states/

Brooks, R., & Grotz, C. (2010). Implementation of electronic medical records: How healthcare providers are managing the challenges of going digital. *Journal of Business & Economics Research, 8*(6), 73–84.

BusinessDictionary.com. (n.d.). Retrieved August 15, 2016, from fragmented market: http://www.businessdictionary.com/definition/fragmented-market.html

Canada Health Infoway Inc. (2012). *Cloud computing in healthcare.* Montreal, Toronto, Vancouver, Halifax: Infoway – Emerging Technology Group (ETG).

Carbon Trust. (2010). *Hospitals.* London: Carbon Trust.

CII – Confederation of Indian Industry. (2010). *Green buildings & green hospitals.* Kochi, CII (Confederation of Indian Industries, India).

Clark, J. (2013). *IT now 10 percent of world's electricity consumption, report finds.* San Francisco: The Register. Retrieved from https://www.theregister.co.uk/2013/08/16/it_electricity_use_worse_than_you_thought/

Cohen, G. (2013). Telemedicine. In G. Cohen (Ed.), *Cohen, Glenn* (pp. 342–356). New York: Oxford University Press.

Collins, N., Novotny, N. L., & Light, A. (2006). A cross-section of readability of health information portability and accountability act authorizations required with health care research. *Journal of Allied Health, 35*, 223–225.

Deloitte. (2012). *Conference background note innovative and sustainable healthcare management: Strategies for growth.* Delhi, India: AIMA (All India Management Association).

Deng, M., Nalin, M., Petković, M., Baroni, I., Marco, A. (2012). Towards trustworthy health platform cloud. *Secure Data Management – Lecture Notes in Computer Science*, pp. 162–175.

Dietrich, B., Plachy, E., & Norton, M. (2014). *Analytics across the enterprise.* Boston: IBM Press. ISBN:978-0-13-383303-4.

Dinodia Capital Advisors. (2012). *Indian healthcare industry.* New Delhi, India: Dinodia Capital Advisors.

Doeksen, G. A., & Schott, V. (2003). Economic importance of the health-care sector in a rural economy. *The International Electronic Journal of Rural and Remote Health Research, Education, Practice and Policy, 3*, 135.

Ebbers, M., Galea, A., Schaefer, M., & Duy Khiem, M. T. (2008). *The green data center: Steps for the journey.* IBM Redpaper. http://www.labouseur.com/courses/di/resources/m06-IBMred-GreenDataCenterJourney.pdf mentions that IBM Red Books are available at the website: https://www.ibm.com/redbooks

Elzen, B., Geels, F. W., & Green, K. (2004). *System innovation and the transition to sustainability: Theory, evidence and policy.* Cheltenham, UK: Edward Elgar Publishing Ltd..

Energy Information Administration (EIA). (2006). *Energy in healthcare fact sheet accompaniment to targeting 100 research study.* Washington, DC: Energy Information Administration (EIA).

EnviroMason. (2016). Retrieved November 12, 2016, from Virginia Mason: https://www.virginia-mason.org/conserving-resources

EPA. (2007). *Report to congress on server and data center energy efficiency – public law 109–431.* Washington, DC: Environmental Protection Agency. Retrieved from http://www.energystar.gov/index.cfm?c=prod_development.server_efficiency_study

EU Code of Conduct. (2017). *2017 best practice guidelines for the EU code of conduct on data centre energy efficiency.* Retrieved from 2017 Best Practice Guidelines for the EU Code of Conduct on Data Centre Energy Efficiency: https://e3p.jrc.ec.europa.eu/publications/2017-best-practice-guidelines-eu-code-conduct-data-centre-energy-efficiency

Franchetti, M. J. (2013). *Carbon footprint analysis* (1st ed.). Boca Raton, FL: CRC Press. ISBN:978-1-4398-5783-0.

Frenk Julio, M. M. (2004). *Health and the economy: A vital relationship.* Retrieved September 24, 2016, from OECD Observer: http://oecdobserver.org/news/archivestory.php/aid/1241/Health_and_the_economy:_A_vital_relationship_.html

Friedman, T. (2008). *Hot, flat, and crowded: Why we need a green revolution – and how it can renew America.* Farrar, MO: Straus and Giroux.

Fries, J. F., Koop, E. C., Beadle, C. E., Cooper, P. P., England, M. J., Greaves, R. F., … Wright, D. (1993). Reducing health care costs by reducing the need and demand for medical services. *The New England Journal of Medicine, 329*, 321–325.

Gail, V. (2002). Green and healthy buildings for the healthcare industry. In *Green and healthy buildings for the Healthcare Industry White Paper*, 10.

Garber, A. M. (2000). Chapter 4 – Advances in cost-effectiveness analysis of health interventions*. In A. M. Garber (Ed.), *Handbook of health economics* (pp. 181–221). New York: Elsevier.

Garber, A. M., & Phelpsc, C. E. (1997). Economic foundations of cost-effectiveness analysis. *Journal of Health Economics, 16,* 1–31.

Gedda, R. (2011, June 9). *Green IT winning favour with CIOs: Survey.* Retrieved April 20, 2014, from CIO: http://www.cio.com.au/article/389632/green_it_winning_favour_cios_survey/

Godbole, N. (2009). *Information systems security: Security management, metrics, frameworks and best practices.* New Delhi, India: Wiley India. ISBN:978-81-265-1692-6.

Godbole, N. (2011a). Green health. In I. R. (USA) (Ed.), *Green technologies* (p. 10). Hershey, PA: IRMA International.

Godbole, N. (2011b). Green health: The green IT implications for healthcare & related businesses. In I. Global (Ed.), *Handbook of research on green ICT: Technology, business and social perspectives* (p. 10). Hershey, PA: IGI Global.

Godbole, N., & Lamb, J. (2013). *The triple challenge for the healthcare industry: Sustainability, privacy, and cloud-centric regulatory compliance, 10th International Conference and Expo on Emerging Technologies for a Smarter World (CEWIT), 2013* (pp. 1–6). Long Island, NY: IEEE.

Godbole, N., & Lamb, J. (2014). *Calculating a hospital's IT energy efficiency and determining cost effective ways for improvement, 11th International Conference & Expo on Emerging Technologies for a Smarter World (CEWIT)* (p. 6). Long Island, NY: IEEE.

Godbole, N., & Lamb, J. (2015). *Using data science and big data analysis to make healthcare green, 12th International Conference & Expo on Emerging Technologies for a Smarter World.* Long Island, NY: IEEE.

Gorbett, T., Salvaterra, D., Skiba, K. (2005, December 2). *Understanding the business rationale behind the trend towards environmentally friendly manufacturing practices.* Williamsburg, VA.

Gore, A. (2007). *An inconvenient truth: The crisis of global warming.* New York: Penguin.

Gore, A. (2017). *An inconvenient sequel: Truth to power.* Emmaus, PA: Rodale Books.

Govindarajan, V., & Ramamurti, R. (2013). Delivering world-class healthcare affordably. *Harvard Business Review (HBR) November 2013, 1–7.*

Grand View Research, Inc. (2016). *Teleradiology market analysis by product (X-ray, computed tomography, ultrasound, nuclear imaging, magnetic resonance imaging) and segment forecasts to 2024.* San Francisco: Grand View Research.

Greenpeace International. (2010, March). *Make IT green: Cloud computing and its contribution to climate change.* Amsterdam: Greenpeace International.

Greenwich Hospital. (2014–2016). *Greenwich hospital.* Retrieved November 12, 2016, from http://www.neep.org/case-study/greenwich-hospital

Gritzalis, S., Lambrinoudakis, C., Lekkas, D., & Deftereos, S. (2005). *Technical guidelines for enhancing privacy and data protection in modern electronic medical environments* (pp. 413–423). Hawaii, USA: IEEE.

Gyan Research and Analytics. (2012). *Indian healthcare industry.* New Delhi, India: Gyan Research and Analytics.

Harris, J. (2008). *Green IT and green computing best practices.* New York: Jason Harris.

Health care accounts for eight percent of U.S. carbon footprint. (2009). Retrieved September 25, 2016, from The University of Chicago Medicine: http://www.uchospitals.edu/news/2009/20091110-footprint.html

Health Research & Educational Trust. (2014). *Environmental sustainability in hospitals: The value of efficiency.* Chicago: HRET (Health Research & Education Trust).

Healthcare without Harm. (2002, January 30). *www.healthybuilding.net.* Retrieved September 25, 2016, from Healthy Building Network and Health Care Without Harm: http://www.healthy-building.net/healthcare/Green_Building_Priorities.pdf

Holmner, Å., Ebi, K. L., Lazuardi, L., & Nilsson, M. (2014). *PLoS One, 9*(9), e105040. https://doi.org/10.1371/journal.pone.0105040. Retrieved May 20, 2017, from plos.org.

Horowitz, B. T. (2012). *Cloud computing in health care to reach $5.4 billion by 2017: Report.* Foster City, CA: eWeek.

Hussain, M. N., & Subramoniam, S. (2012). *Greener healthcare using ICT based BPR, International Conference on Green Technologies (ICGT), 2012* (pp. 215–222). Trivandrum, India: IEEE.

Indian Telemedicine Network. (n.d.). *Introduction to telemedicine.* Retrieved October 27, 2016, from Indian Telemedicine Network: http://telemedindia.org/telemedicine.html

Information Resources Management Association (USA). (2011). In N. Godbole (Ed.), *Green technologies: Concepts, methodologies, tools and applications* (p. 10). Hershey, PA: Information Resources Management Association (IRMA International).

JAMA Editor. (2009). Estimate of the carbon footprint of the US health care sector. *The Journal of American Medical Association (JAMA), 302,* 1970–1972.

Jaswanth, N., Durga, J., & Kmar, K. D. (2013). Migrating health care by analysing MYTHS in cloud technology. *Oriental Journal of Computer Science & Technology, 6,* 49–54.

J-Ath-Training. (2004, January–March). Evidence-based medicine: What is it and how does it apply to athletic training? *Journal of Athletic Training, 39,* 83–87.

Johnson & Johnson. (2012). The growing importance of more sustainable products in the global health care industry. Retrieved from https://www.jnj.com/_document?id=00000159-6a81-dba3-afdb-7aeba25f0000

Judge, W. Q., & Ryman, J. A. (2001). The shared leadership challenge in strategic alliances: Lessons from the U.S. healthcare industry. *Academy of Management, 15,* 71–79.

Khalil Khoumbati, Y. K. (2010). *Handbook of research on advances in health informatics and electronic healthcare applications: Global adoption and impact of information communication technologies.* Hershey, PA: IGI Global.

Kim, S. (N.A.). *Green IT best practices.* Washington, DC: American Hotel and Lodging Association. Retrieved from https://www.google.co.in/?gws_rd=cr&ei=vfPYUtaOMeTkiAe1_YBo#q=green+it+definition&start=10

Kizer, K. W. (1999). The "New VA": A national laboratory for health care quality management. *American Journal of Medical Quality, 14,* 3–20.

Koomey, J. (2007). *Data center electricity use.* Santa Clara, CA: Lawrence Berkeley National Laboratory & Stanford University.

KPMG. (2011). *Increasing importance of social media in healthcare.* Amsterdam: KPMG International.

Lamb, J. (2009). *The greening of IT: How companies can make a difference for the environment.* Boston: IBM Press. ISBN:978-0-13-715083-0.

Lamb, J. (2011). *Green IT and use of private cloud computing in South Africa, CEWIT 2011 conference.* Long Island, USA: IEEE Xplore.

Laranjo, L., Lau, A., Oldenburg, B., Gabarron, E., O'Neill, A., Chan, S., & Coiera, E. (2015). *mHealth technologies for chronic disease prevention and management.* Sydney, Australia: Sax Institute.

Leetz, A. (2011). *Transforming healthcare for the 21st century.* Healthcare without Harm. http://www.who.int/globalchange/publications/climatefootprint_report.pdf

Liu, L., Wang, H., Liu, X., Jin, X., He, W. B., Wang, Q. B., & Chen, Y. (2012). GreenCloud: A new architecture for green data center. *The Journal of Supercomputing, 60,* 268–280.

Lunde, S. (2013). *The mhealTh case in india: Telco-led transformation of healthcare service delivery in India.* Bangalore, India: Wipro Council for Industry Research.

Mahmud Adeb, P. M. (2007). *The role of the healthcare sector in expanding economic opportunity.* Cambridge, MA: Harvard University.

Marston, S., Li, Z., Bandyopadhyay, S., Zhang, J., & Ghalsasi, A. (2011). Cloud computing – the business perspective. *Decision Support Systems (Elsevier), 51,* 176–189.

Masino, C., Rubinstein, E., Lem, L., Purdy, B., & Rossos, P. G. (2010, November). The impact of telemedicine on greenhouse gas emissions at an academic health science center in Canada. *Telemedicine and e-Health, 16,* 973–976.

Mechael, P., Batavia, H., Kaonga, N., Searle, S., Kwan, A., Goldberger, A., … Ossman, J. (2010). *Barriers and gaps affecting.* New York: The Earth Institute, Colombia University, mHealth Alliance.

Medeiros de Bustos, E., & Moulin, T. (2009). Barriers, legal issues, limitations and ongoing questions in telemedicine applied to stroke. *The Journal of Cerebrovascular Diseases, 4*, 36–39.

mHealth Case in India. (n.d.). Telco-led transformation of healthcare service delivery in India the report available at the URL: http://www.wipro.com/documents/the-mHealth-Case-in-India.pdf, Accessed on 9th May 2018.

Mickan, S., Tilson, J., Atherton, H., Roberts, N. W., & Heneghan, C. (2013, October 28). Evidence of effectiveness of health care professionals using handheld computers: A scoping review of systematic reviews. *Journal of Medical Internet Research, 15*(10), e212.

Microsoft and Accenture. (2010). *Microsoft, accenture and WSP environment & energy study shows significant energy and carbon emissions reduction potential from cloud computing.* Microsoft and Accenture.

Mines, C. (2011). *4 reasons why cloud computing is also a green solution.* Retrieved April 29, 2014, from GreenBiz.com: http://www.greenbiz.com/blog/2011/07/27/4-reasons-why-cloud-computing-also-green-solution

Mingay, S. (2007). *Green IT: The new industry shock wave.* Stamford, CT: Gartner.

Muduli, K. A. (2012, August). Barriers to green practices in health care waste sector: An Indian perspective. *International Journal of Environmental Science and Development, 3*(4), 393–399.

Murugesan, S., & Gangadharan, G. R. (2012). *Harnessing green IT: Principles and practices.* London: Wiley – IEEE.

O'Donnell, S. (2014). *Is cloud computing more energy efficient than traditional computing?* Retrieved May 3, 2014, from Ask – Q&A (Computers and Electronics): http://www.ask.com/question/is-cloud-computing-more-energy-efficient-than-traditional-computing

Obama, B. (2017). The irreversible momentum of clean energy. *Science Magazine.*

OECD – Organisation for Economic Co-operation and Development. (2013). *Health at a glance 2013: OECD indicators.* Paris: OECD Publishing.

Padhy, R. P., Patras, M. R., & Satapathy, S. C. (2012). *Design and implementation of a cloud based rural healthcare information system model.* Bangalore, India: Oracle Corporation.

Pope-Francis. (2015). *Laudato Si – On care for our common home.* Washington, DC: USCCB Communications.

Porter, M. E. (2007, October 4). *Value-based health care delivery: Implications for providers, Tosteson Lecture – Harvard Medical School.* Boston, MA: Harvard Business School.

PR Newwire. (2012). *Data center construction boom driven by healthcare and technology.* Retrieved April 10, 2014, from PR Newswire: http://www.prnewswire.com/news-releases/data-center-construction-boom-driven-by-healthcare-and-technology-138810069.html

Practice Greenhealth. (2014). *Advancing sustainability in health care: A collection of special case studies.* Reston, VA: Practice Greenhealth.

Premier, Inc. (2013). *Healthcare energy reduction & efficiency.* Retrieved January 18, 2014, from Transforming Healthcare Together, Premier Inc.: https://www.premierinc.com/quality-safety/tools-services/safety/topics/energy/index.jsp

Preuveneersa, D., Berbers, Y., & Joosen, W. (2013). The future of mobile E-health application development: Exploring HTML5 for context-aware diabetes monitoring. *Procedia Computer Science, 21*, 351–359.

Protected Healthcare Information (PHI). (n.d.). Retrieved from https://www.hhs.gov/answers/hipaa/what-is-phi/index.html

Provost, F., & Fawcett, T. (2013). *Data science for business – what you need to know about data mining and data analytic thinking.* Sebastopol, CA: O'Reilly. ISBN:978-1-449-36132-7.

Ramakrishnan, S. (2008). *Health informatics in India: Vision and activities, Symoposium on Medical Informatics in Indian Context and Unveiling of C-DAC's Medical Informatics Standard SDK Suite* (p. 27). Delhi, India: Center of Development of Advanced Computing (C-DAC).

Rebecca, H. S. (2015, February 4). *2015 hospital construction survey.* Retrieved September 24, 2016, from Health Facilities Management: http://www.hfmmagazine.com/articles/1474-hospital-construction-survey-results-are-in

Ren, Y., Werner, R., Pazzi, N., & Boukerche, A. (2010). Monitoring patients via a secure and mobile healthcare system. *IEEE, 17*, 59–65.

Roberts, R. R., Frutos, P. W., Ciavarella, G. G., Gussow, L. M., Mensah, E. K., Kampe, L. M., … Rydman, R. J. (1999). Distribution of variable vs fixed costs of hospital care. *The Journal of American Medical Association, 281*, 644–649.

Santoro, E., Castelnuovo, G., Zoppis, I., Mauri, G., & Sicurello, F. (2015, May 7). Social media and mobile applications in chronic disease prevention and management. *Frontiers in Psychology, 7*, 567.

Scherr, D., Kastner, P., Kollmann, A., Auer, J., Krappinger, H., Schuchlenz, H., … Investigators, M. (2009). Effect of home-based telemonitoring using mobile phone technology on the outcome of heart failure patients after an episode of acute decompensation: Randomized controlled trial. *Journal of Medical Internet Research (JMR), 11*(3), e34.

Shoui, Y. (2013). *Perspectives on supply chain management in the healthcare industry, International conference on science and social research* (pp. 630–633). Beijing, China: ICSSR.

Siebenaller, B. (2012). *Connecting sustainability to the healthcare mission.* Retrieved May 19, 2014, from Healthcare Design: http://www.healthcaredesignmagazine.com/article/connecting-sustainability-healthcare-mission

Singer, B. C., & Tschudi, W. F. (2009). *High performance healthcare buildings: A roadmap to improved energy efficiency.* Berkeley, CA: Ernest Orlando Lawrence Berkeley National Laboratory.

Singer, B. C., Coughlin, J. L., Mathew, P. A. (2009). *Summary of information and resources related to energy use in healthcare facilities – version 1.* Retrieved October 7, 2013, from http://hight-ech.lbl.gov/documents/healthcare/lbnl-2744e.pdf

Steve, R. (2012, February 9). *Healthcare facilities survey indicates growing concerns about 'Green Washing'.* Retrieved September 24, 2016, from ConstructionPro Network: http://construction-pronet.com/Content_Free/2912part2.aspx

Sullivan, C. (2009, July 8). *Going green in healthcare IT and improving organizational efficiency along the way.* Retrieved January 21, 2013, from HealthBlog: http://blogs.msdn.com/b/health-blog/archive/2009/07/08/going-green-in-healthcare-it-and-improving-organizational-effi-ciency-along-the-way.aspx

Sullivan, M. (2012, March). Egging energy efficiency for profit. *Express Healthcare.*

Sun, J., Reddy, C. (2013). Big data analytics for healthcare, tutorial presentation. In *SIAM International Conference on Data Mining.* Austin, TX.

Swan, G. (2011, June 9). *How green is my cloud?* Retrieved April 2014 20, 2014, from CIO: http://www.cio.com.au/article/389633/how_green_my_cloud_/

TechTarget. (2014). Green computing energy efficiency guide.

The 2011 IBM Tech Trends Report, November 15, 2011, ibm.com/developerworks/techtrendsreport available also at the URL https://ai.arizona.edu/sites/ai/files/MIS510/2011ibmtechtrendsreport.pdf, accessed on 5th May, 2018.

The Economist. (2016, April 16). The new sunbathers. *The Economist*, p. 12.

The Green Grid. (n.d.). *The green grid data center power efficiency.* Retrieved from http://www.thegreengrid.org

The World Economic Forum's Insight Report. (2012). The Global Competitiveness Report 2012–2013. Accessed on 19th May 2018 at the URL: http://www3.weforum.org/docs/WEF_GlobalCompetitivenessReport_2012-13.pdf.

UC Hospitals. (2009). *Healthcare accounts for eight percent of U.S. Carbon Footprint.* UC Hospitals. https://www.sciencedaily.com/releases/2009/11/091110171647.htm

US DOE Annual Energy Outlook. (2017). https://www.eia.gov/outlooks/aeo/pdf/electricity_gen-eration.pdf

Varon, E. (2007, March 28). *Why green IT is better IT.* Retrieved April 20, 2014, from CIO (The CIO Magazine): http://www.cio.com/article/100557/Why_Green_IT_is_Better_IT

Walsh, B. (2008, July 18). Gore's bold, unrealistic plan to save the planet. *Time Magazine.* http://www.time.com/time/health/article/0,8599,1824132,00.html?xid=newsletter-weekly

Weisbrod, B. A. (1991). The health care quadrilemma: An essay on technological change, insur-ance, quality of care, and cost containment. *Journal of Economic Literature, 29*, 523–552.

West, D. M. (2010). *Overcoming rural health care barriers through innovative wireless health technologies*. Retrieved October 27, 2016, from Brookings: https://www.brookings.edu/testimonies/overcoming-rural-health-care-barriers-through-innovative-wireless-health-technologies/

West, D. (2012a). How mobile devices are transforming healthcare. *Issues in Technology Innovation, 18*, 1–11.

West, D. M. (2012b). *How mobile devices are transforming healthcare*. Washington, DC: Center for Technology Innovation at the Brookings Institution.

Wikipedia. (2016, August 31). *mHealth*. Retrieved October 23, 2016, from Wikipedia, page was last modified on 31 August 2016, at 16:11.: https://en.wikipedia.org/wiki/MHealth

Wikipedia. (2017). *Personal health record (PHR)*. Retrieved from Wikipedia: http://en.wikipedia.org/wiki/Personal_health_record

Wikipedia. (n.d.). *Wikipedia*. Retrieved from Wikipedia: https://en.wikipedia.org/wiki/Health_care

World Economics Forum. (2013). *Sustainable health systems: visions, strategies, critical uncertainties and scenarios*. Geneva, Switzerland: World Health Organization.

Yanos, B., White, E., Suprina, D., Parker, J., Sutton, R. (2009). Healthcare clients adopt electronic health records with cloud-based services. *Cloud Computing Journal*.

Yellowlees, P. M., Chorba, K., Parish, M. B., Wynn-Jones, H., & Nafiz, N. (2010). Telemedicine can make healthcare greener. *Telemedicine and e-Health, 16*, 229–232.

Index

© Springer International Publishing AG, part of Springer Nature 2018
N. S. Godbole, J. P. Lamb, *Making Healthcare Green*,
https://doi.org/10.1007/978-3-319-79069-5

Printed in the United States
By Bookmasters